The Murders in the Rue Morgue
and The Purloined Letter

莫爾格街兇殺案

U0109096

商務印書館

Name of Book: The Murders in the Rue Morgue and The Purloined Letter
Author: Edgar Allan Poe
Text adaptation and activities: Graeme and Silvia Thomson
Editors: Rebecca Raynes, Elvira Poggi Repetto
Design: Nadia Maestri
Illustrations: Gianni De Conno
Edition: ©2000 Black Cat Publishing
                 an imprint of Cideb Editrice, Genoa, Canterbury

系 列 名：Black Cat 優質英語階梯閱讀 · Level 6
書　　名：莫爾格街兇殺案
責任編輯：黃淑嫻
封面設計：張　毅
出　　版：商務印書館（香港）有限公司
　　　　　香港筲箕灣耀興道3號東滙廣場8樓
　　　　　http://www.commercialpress.com.hk
印　　刷：中華商務彩色印刷有限公司
　　　　　香港新界大埔汀麗路36號中華商務印刷大廈
版　　次：2004年2月第1版第1次印刷
　　　　　© 2004 商務印書館（香港）有限公司
　　　　　ISBN 962 07 1700 7
　　　　　Printed in Hong Kong

出版説明

　　本館一向倡導優質閱讀，近年來連續推出了以"Q"為標識的
"Quality English Learning 優質英語學習"系列，其中《讀名著學英語》叢
書，更是香港書展入選好書，讀者反響令人鼓舞。推動社會閱讀風氣，推
動英語經典閱讀，藉閱讀拓廣世界視野，提高英語水平，已經成為一種潮
流。

　　然良好閱讀習慣的養成非一日之功，大多數初、中級程度的讀者，常
視直接閱讀厚重的原著為畏途。如何給年輕的讀者提供切實的指引和幫
助，如何既提供優質的學習素材，又提供名師的教學方法，是當下社會關
注的重要問題。針對這種情況，本館特別延請香港名校名師，根據多年豐
富的教學經驗，精選海外適合初、中級英語程度讀者的優質經典讀物，有
系統地出版了這套叢書，名為《Black Cat 優質英語階梯閱讀》。

　　《Black Cat 優質英語階梯閱讀》體現了香港名校名師堅持經典學習的
教學理念，以及多年行之有效的學習方法。既有經過改寫和縮寫的經典名
著，又有富創意的現代作品；既有精心設計的聽、説、讀、寫綜合練習，
又有豐富的歷史文化知識；既有彩色插圖、繪圖和照片，又有英美專業演
員朗讀作品的 CD。適合口味不同的讀者享受閱讀之樂，欣賞經典之美。

　　《Black Cat 優質英語階梯閱讀》由淺入深，逐階提升，好像參與一個
尋寶遊戲，入門並不難，但要真正尋得寶藏，需要投入，更需要堅持。只
有置身其中的人，才能體味純正英語的魅力，領略得到真寶的快樂。當英
語閱讀成為自己生活的一部分，英語水平的提高自然水到渠成。

商務印書館 (香港) 有限公司
編輯部

使用説明

① 應該怎樣選書？

按閱讀興趣選書

《Black Cat 優質英語階梯閱讀》精選世界經典作品，也包括富於創意的現代作品；既有膾炙人口的小說、戲劇，又有非小說類的文化知識讀物，品種豐富，內容多樣，適合口味不同的讀者挑選自己感興趣的書，享受閱讀的樂趣。

按英語程度選書

《Black Cat 優質英語階梯閱讀》現設 Level 1 至 Level 6，由淺入深，涵蓋初、中級英語程度。讀物分級採用了國際上通用的劃分標準，主要以詞彙（vocabulary）和結構（structures）劃分。

Level 1 至 Level 3 出現的詞彙較淺顯，相對深的核心詞彙均配上中文解釋，節省讀者查找詞典的時間，以專心理解正文內容。在註釋的幫助下，讀者若能流暢地閱讀正文內容，就不用擔心這一本書程度過深。

Level 1 至 Level 3 出現的動詞時態形式和句子結構比較簡單。動詞時態形式以現在時（present simple）、現在時進行式（present continuous）、過去時（past simple）為主，句子結構大部分是簡單句（simple sentences）。此外，還包括比較級和最高級（comparative and superlative forms）、可數和不可數名詞（countable and uncountable nouns）以及冠詞（articles）等語法知識點。

Level 4 至 Level 6 出現的動詞時態形式，以現在完成時（present perfect）、現在完成時進行式（present perfect continuous）、過去完成時（past perfect continuous）為主，句子結構大部分是複合句（compound sentences）、條件從句（1st and 2nd conditional sentences）等。此外，還包括情態動詞（modal verbs）、被動形式（passive forms）、動名詞（gerunds）、

短語動詞（phrasal verbs）等語法知識點。

　　根據上述的語法範圍，讀者可按自己實際的英語水平，如詞彙量、語法知識、理解能力、閱讀能力等自主選擇，不再受制於學校年級劃分或學歷高低的約束，完全根據個人需要選擇合適的讀物。

② 怎樣提高閱讀效果？

　　閱讀的方法主要有兩種：一是泛讀，二是精讀。兩者各有功能，適當地結合使用，相輔相成，有事半功倍之效。

　　泛讀，指閱讀大量適合自己程度（可稍淺，但不能過深）、不同內容、風格、體裁的讀物，但求明白內容大意，不用花費太多時間鑽研細節，主要作用是多接觸英語，減輕對它的生疏感，鞏固以前所學過的英語，讓腦子在潛意識中吸收詞彙用法、語法結構等。

　　精讀，指小心認真地閱讀內容精彩、組織有條理、遣詞造句又正確的作品，着重點在於理解"準確"及"深入"，欣賞其精彩獨到之處。精讀時，可充分利用書中精心設計的練習，學習掌握有用的英語詞彙和語法知識。精讀後，可再花十分鐘朗讀其中一小段有趣的文字，邊唸邊細心領會文字的結構和意思。

　　《Black Cat 優質英語階梯閱讀》中的作品均值得精讀，如時間有限，不妨嘗試每兩個星期泛讀一本，輔以每星期挑選書中一章精彩的文字精讀。要學好英語，持之以恆地泛讀和精讀英文是最有效的方法。

③ 本系列的練習與測試有何功能？

　　《Black Cat 優質英語階梯閱讀》特別注重練習的設計，為讀者考慮周到，切合實用需求，學習功能強。每章後均配有訓練聽、説、讀、寫四項技能的練習，分量、難度恰到好處。

聽力練習分兩類，一是重聽故事回答問題，二是聆聽主角對話、書信朗讀、或模擬記者訪問後寫出答案，旨在以生活化的練習形式逐步提高聽力。每本書均配有 CD 提供作品朗讀，朗讀者都是專業演員，英國作品由英國演員錄音，美國作品由美國演員錄音，務求增加聆聽的真實感和感染力。多聆聽英式和美式英語兩種發音，可讓讀者熟悉二者的差異，逐漸培養分辨英美發音的能力，提高聆聽理解的準確度。此外，模仿錄音朗讀故事或模仿主人翁在戲劇中的對白，都是訓練口語能力的好方法。

閱讀理解練習形式多樣化，有縱橫字謎、配對、填空、字句重組等等，注重訓練讀者的理解、推敲和聯想等多種閱讀技能。

寫作練習尤具新意，教讀者使用網式圖示（spidergrams）記錄重點，採用問答、書信、電報、記者採訪等多樣化形式，鼓勵讀者動手寫作。

書後更設有升級測試（Exit Test）及答案，供讀者檢查學習效果。充分利用書中的練習和測試，可全面提升聽、說、讀、寫四項技能。

❹ 本系列還能提供甚麼幫助？

《Black Cat 優質英語階梯閱讀》提倡豐富多元的現代閱讀，巧用書中提供的資訊，有助於提升英語理解力，擴闊視野。

每本書都設有專章介紹相關的歷史文化知識，經典名著更有作者生平、社會背景等資訊。書內富有表現力的彩色插圖、繪圖和照片，使閱讀充滿趣味，部分加上如何解讀古典名畫的指導，增長見識。有的書還提供一些與主題相關的網址，比如關於不同國家的節慶源流的網址，讓讀者多利用網上資源增進知識。

Contents

The Purloined Letter
失竊的信

The Murders in the Rue Morgue is recorded in full on the CD. 故事選錄

These symbols indicate the beginning and end of the extracts linked to the listening activities. 聽力練習開始和結束的標記

Some Information about
Edgar Allan Poe's Life

Edgar Allan Poe was born in Boston in 1809. Both his parents, David Poe and Elizabeth Arnold, were itinerant [1] actors and died of consumption [2] when he was very young. Although never officially adopted, Poe was taken into the home of John Allan, a merchant from Virginia, and his wife, Frances Keeling Valentine to whom the writer became particularly attached.

In 1815 the Allans moved to Britain where the family lived for five years and where Edgar also went to school. On their return to Virginia, Poe was sent to a private school where he was considered an exceptional all-round student.

In 1826 Poe enrolled at the University of Virginia but his foster-[3]father refused to finance his studies. This made their relationship even more difficult and Edgar was forced to leave the University after only one

1. **itinerant** : travelling.
2. **consumption** : (dated) tuberculosis, a disease of the lungs.
3. **foster-** : with a family connection through taking care of and bringing up rather than of birth.

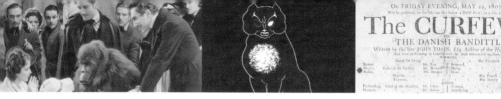

year, despite his excellent results. In the same period Poe started to gamble and drink very heavily. A year later, he moved to Boston where he began his literary career in earnest and suffered his first disappointment when *Tamerlane and Other Poems by a Bostonian,* which he published himself, was totally ignored by critics.

In 1827 Edgar enlisted [1] in the Army under the name Edgar A. Perry.

His quarrels with John Allan continued. In 1829 after returning to Richmond too late to attend his foster-mother's funeral, he decided to apply for a cadetship at West Point. [2] Before entering West

John Allan.

Point, Edgar submitted a manuscript entitled *Al Aaraaf, Tamerlane, and Minor Poems* for publication. This time the book was published not anonymously, [3] but under the name Edgar A. Poe, the middle initial acknowledging the part the Allans had played in his life. Expelled [4] from the Academy in 1832, Poe moved to Baltimore to live with his aunt, Maria Clemm, and his first cousin Virginia. The following year he won a literary prize for his story

Maria Clemm.

1. **enlisted** : joined, registered.
2. **West Point** : a military academy.
3. **anonymously** : without the name.
4. **expelled** : forced to leave.

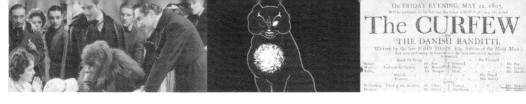

MS. Found in a Bottle in the *Baltimore Saturday Visiter*. In 1835 Poe brought his aunt and cousin to Richmond where he worked with Thomas Willis White at the *Southern Literary Messenger*. The next year he married his cousin Virginia, only thirteen years old. His work at the *Messenger* was mainly criticism such as reviews of other magazines, novels and poetry and helped to boost [1] the magazine's sales.

His editorial work, however, was not sufficient to support his family, and the Poes had trouble making ends meet. [2] In 1840 Poe published *Tales of the Grotesque and Arabesque* which contains several of his greatest stories, followed a year later by *The Murders in the Rue Morgue* which is considered by some critics to be the first ever detective story. However it was *The Gold Bug*, published in 1843, which brought Poe widespread acclaim. [3]

Financial success soon followed with the poem *The Raven* which first appeared in *The Evening Mirror*. The poem is a perfect example of Poe's mastery of rhythm and lyrical invention.

Tragedy befell [4] Poe in 1847 when his young wife Virginia died of tuberculosis. The heartbroken author began to drink more heavily. Writing of the effect of Virginia's death, Poe remarked: "I became insane, [5] with long intervals of horrible sanity. [6] During these fits of absolute unconsciousness, I drank . . . my enemies referred the insanity to the drink, rather than the drink to the insanity."

On October 3, 1849 he was found unconscious in front of a polling

---

1. **boost** : increase, improve.
2. **making ends meet** : managing financially.
3. **acclaim** : public praise.
4. **befell** : happened to.
5. **insane** : mentally ill.
6. **sanity** : health of mind.

CHAPTER 1

Extraordinary[1] Murders

I was staying in Paris during the spring and part of the summer of 18__. There I met a Monsieur[2] C. Auguste Dupin. This young gentleman came from a noble family but he himself was not very rich. He was not really interested in money. He lived frugally.[3] Books were his only luxury.

We first met at an obscure[4] library in the Rue Montmartre. By some strange coincidence[5] we were both looking for the same book. After that, we met many more times.

He told me about the history of his family. I was astonished[6]

1. **extraordinary** : very unusual.
2. **Monsieur** : (French) Mr.
3. **frugally** : without spending much money.
4. **obscure** : not well-known.
5. **coincidence** : chance.
6. **astonished** : very surprised.

by the extent of his reading. When did he find the time to read so many books? And I was fascinated by his vivid [1] imagination.

It was decided that we should live together while I was in Paris. We found a big, old, deserted [2] house in the Faubourg St. Germain. As I had more money than Dupin, I offered to pay the rent.

Our isolation was perfect. We admitted no visitors. Nobody knew the address of the house where we lived.

For some strange reason, Dupin loved the night and I began to share his enthusiasm. Of course the night did not last forever, so when the morning came we closed all the shutters [3] on the building to simulate the conditions of darkness. Then we lit two or three candles. This was to have enough light to read or write or simply talk. We sat in the house all day until the clock indicated that the true night was coming. Then we went out into the streets continuing our conversation. We walked far and wide in the great city, looking for things to stimulate our imagination. There was an infinity [4] of mental excitement in simply observing the world.

It was during these walks that I discovered and began to admire Dupin's incredible

1. **vivid** : lifelike.
2. **deserted** : empty and quiet because no people there.
3. **shutters** : covers on windows to keep out the light.
4. **infinity** : state of being endless.

analytic ability. He told me that he could see directly into men's hearts and minds. At first I did not believe him. But then something happened to change my mind. One night we were walking down a long dirty street near the Palais Royal. I was looking at a newspaper when I noticed one particular story:

EXTRAORDINARY MURDERS

This morning at about three o'clock, a number of terrible screams awoke the inhabitants [1] of the Quartier St. Roch. The screams were coming from the fourth floor of a house in the Rue Morgue. Only two people lived there: Madame [2] L'Espanaye and her daughter Mademoiselle [3] Camille L'Espanaye. After several attempts, neighbours finally entered the house together with two policemen. By this time there were no more screams. But as the group ran up the stairs, they heard two more angry voices coming from the upper part of the house. However, when they got to the fourth floor there was again silence. They divided up into small groups, moving from room to room. They finally arrived at a back room. The scene they discovered there was almost too horrible to describe.

The apartment was in great disorder. The furniture was broken and the bed lay in the middle of the floor. On a chair was a razor covered in blood. Bloody lengths of human hair lay in the fireplace. On the floor there were three large silver spoons, an ear-ring and two bags containing four thousand francs in gold. The

1. **inhabitants** : people that live in a particular place.
2. **madame** : (French) Mrs.
3. **mademoiselle** : (French) Miss.

The Murders in the Rue Morgue

drawers of a desk which stood in one corner were open and papers were scattered about. [1]

Under the bed was an open safe with the key still in the door. It contained a few letters and other papers but nothing of any importance.

There was no sign of Madame L'Espanaye. But somebody noticed there was an extraordinary quantity of soot [2] in the fireplace. And so they searched the chimney. There they found the dead body of the daughter. The body was quite warm. A doctor examined it and found many bruises [3] and cuts. On the throat there were several dark bruises and finger marks. This suggested only one thing. Strangulation. [4]

They searched the rest of the house but found nothing. Finally they went into the small garden at the back of the house. There they found the body of the old lady. Her throat was completely cut and when two men tried to raise her, the head fell off. The body itself was completely mutilated. [5] It didn't look human.

No one has yet found a clue that could help to solve this mystery.

1. **scattered about** : thrown around untidily.
2. **soot** : black powder which rises in the smoke from a fire.
3. **bruises** : purple marks on the body.
4. **strangulation** : the action of killing someone by pressing one's throat.
5. **mutilated** : severely damaged someone's body by cutting it.

Monsieur C. Auguste Dupin

1 Complete the following sentences to describe Dupin and his peculiar [1] habits. (Try to use your own words!)

a. Although Dupin came from a noble family, ...

b. Books were ....

c. Because of Dupin's financial difficulties, the narrator ....

d. Darkness was ....

e. During the day ....

f. At night ....

g. Thanks to his amazing analytical ability, ...

What happened in Chapter One?

2 Answer the following questions.

a. What were the narrator and Dupin doing when they met?

b. How did they learn about the terrible events that had occurred in Quartier St. Roch?

c. Who were the first people to know of the murders? How?

d. What did the police find in Madame L'Espanaye's apartment?

e. Where did they find the dead body of Madame L'Espanaye's daughter?

f. How did she die?

g. Where was the corpse of the old lady?

h. How was she killed?

i. What was the police's theory regarding these two horrible murders?

---

1. **peculiar** : unusual and strange.

The art of analysis

3 The analytical mind, according to Poe, derives pleasure from even the most trivial [1] occupations and is curious about anything that may bring its talent into play. It is fond of enigmas, [2] puzzles, hieroglyphics, [3] everything concerned with the faculty [4] of intuition. [5]

a. According to you, what are the skills that a good detective needs?

b. Are there any detectives, either in fiction or in films, that you admire in particular?

c. What is their method for solving a crime?

d. Do you consider yourself to have an analytical mind?

e. Would you make a good investigator? Why/Why not?

f. Imagine you are in charge of the murder investigation and you are in Madame L'Espanaye's apartment. From the clues you have, try to solve the following mysteries:

- The apartment is in great disorder. Why?
- In the fireplace you see "bloody lengths of human hair". Who does it belong to?
- How could the corpse of the girl be stuck into the chimney?
- What might the murderer's motive be?

Keep on reading! You will find more evidence later on in the story.

1. **trivial** : having little importance.
2. **enigmas** : mysteries.
3. **hieroglyphics** : a system of writing using pictures instead of words.
4. **faculty** : ability.
5. **intuition** : the ability to understand something because of a feeling.

Tense revision – Edgar Allan Poe's parents

4 Read this text on Edgar Allan Poe's parents and fill in the gaps with the verbs in brackets in the right tense.

Edgar Allan Poe .... *(be born)* on 19 January 1809. His parents, David and Elizabeth Arnold Poe, were professional actors. Elizabeth Poe .... *(acclaim)* by the public in Boston where she .... *(find)*, in her own words, "her best and most sympathetic friends". David Poe .... *(abandon)* the study of law to become an actor but he .... *(not prove)* to be very talented; on the other hand, Poe's mother was not only a skilled actress but a very precocious [1] one. By the time she was fourteen she .... *(already, act)* as Ophelia and at the time of her death at the age of twenty-four her repertoire [2] .... *(include)* over two hundred roles.

However, the Poes .... *(not have)* financial success as actors and they always .... *(lead)* a very precarious [3] life. This explains why, even when Elizabeth was pregnant with Edgar and .... *(struggle)* with all kinds of difficulties, she almost never stopped .... *(act)*.

All records of David Poe .... *(end)* in July 1810. Their first son, William Henry Poe .... *(take)* into the home of his grandparents but Edgar .... *(stay)* with his mother who was pregnant with a third child and already very ill with tuberculosis.

In December 1811 Elizabeth Poe, who .... *(once, be)* a great actress, ... *(die)* in extreme poverty in a Richmond boarding house. William Henry .... *(still, live)* with his grandparents at the time, while Edgar .... *(adopt)*, though never officially, by the merchant John Allan and his wife. Edgar's little sister, the infant Rosalie, ... *(shelter)* by another Richmond family, the William Mackenzies.

1. **precocious** : showing unusually early achievement.
2. **repertoire** : all the plays that actors can perform.
3. **precarious** : uncertain, unsafe.

Passive

5 Turn the following sentences into the passive form.

a. Since Dupin was not particularly well-off, the narrator was paying the rent.

   ....
   ....

b. If the neighbours hadn't heard the voices, nobody would have found out about the murders.

   ....
   ....

c. As the group ran up the stairs, they heard two angry voices coming from the upper part of the house.

   ....
   ....

d. The police discovered a horrible scene in Madame L'Espanaye's flat.

   ....
   ....

e. They had found the girl's body in the chimney.

   ....
   ....

f. No one has yet found a clue that could help to solve the mystery.

   ....
   ....

Before you read

1 Listen to the beginning of Chapter 2 and the information about the first two witnesses, Pauline Dubourg and Pierre Moreau, and fill in the gaps with the missing words.

Pauline Dubourg, Madame L'Espanaye's ...., says that she has known both the .... for three years. The old lady and her .... seemed to have a good relationship. They were very .... to each other. They .... her very well. She didn't know what Madame L'Espanaye's .... was. She never met anyone in the house when she came for the .... They didn't have a .... There was no furniture in the building apart from that in the .... floor apartment where they lived.

Pierre Moreau, ..., says that he has sold tobacco to Madame L'Espanaye for almost four years. He was .... in the area and has always lived there. The victims .... to the house six years ago. The two of them lived a very .... life. He .... they had money. The only people who .... the house were the old lady and her daughter, a .... once or twice and a .... eight or ten times.

Now read the text and check your answers.

23

CHAPTER 2

The Testimonies [1]

 The next day's paper had the following additional details:
The Tragedy of the Rue Morgue.

Police have questioned many individuals about this horrible incident. The truth behind the murders, however, still remains a mystery.

Below we have printed the testimonies of the neighbours and witnesses: [2]

Pauline Dubourg, Madame L'Espanaye's laundress, [3] says that she has known both the victims for three years. The old lady and her daughter seemed to have a good relationship. They were very affectionate [4] to each other. They paid her very well. She didn't

---

1. **testimonies** : spoken or written statements that something is true.
2. **witnesses** : people who see an accident or crime (In this case they didn't see anything, but they heard something.).
3. **laundress** : woman who washes (launders) clothes professionally.
4. **affectionate** : showing feelings of love.

know what Madame L'Espanaye's job was. She never met anyone in
the house when she came for the washing. They didn't have a
servant. There was no furniture in the building apart from that in
the fourth floor apartment where they lived.

Pierre Moreau, tobacconist, says that he has sold tobacco to
Madame L'Espanaye for almost four years. He was born in the area
and has always lived there. The victims moved to the house six
years ago. The two of them lived a very quiet life. He believed they
had money. The only people who entered the house were the old
lady and her daughter, a porter [1] once or twice and a doctor eight or
ten times.

Other neighbours said similar things. There were never any
visitors to the house. Nobody knew if Madame L'Espanaye had any
relatives. The shutters of the front windows were usually closed and
those on the windows at the back of the house were always closed
with one exception: the large room at the back on the fourth floor. It
was a good house and wasn't very old.

Isidore Muset, policeman, says that someone called him and told
him to go to the house. There he found about twenty or thirty
people at the gates. They were trying to get in. He opened the gates
easily with a piece of metal. The screams continued until the gates
were open. Then they stopped. They seemed to be screams of a
person (or people) in great agony. [2] They were loud and long. The

1. **porter** : person whose job is to carry things.
2. **agony** : extreme physical or mental suffering.

party [1] went upstairs. From the first floor they could hear two voices. They seemed to be arguing. One was quite low, the other much higher – a very strange voice. The first voice was that of a Frenchman. Not a woman. The other voice was that of a foreigner [2] but he could not tell if it was a man or a woman. He thought the language was Spanish but Mr Muset does not speak Spanish himself.

Henri Duval, a neighbour, says that he was one of the party who first entered the house. In general he agrees with the testimony of Muset. But he thinks that the high voice was that of an Italian although he does not speak Italian. He is certain it was not French. He could not be sure that it was a man's voice. Possibly a woman. He knew Madame L'Espanaye and her daughter. He was sure that the high voice did not belong to either of them.

Monsieur Odenheimer, restaurant owner, comes from Amsterdam and does not speak French. He was passing the house when he heard the screams. They lasted for about ten minutes. He was one of those who entered the building. But he was sure that the high voice was that of a man – a Frenchman. He didn't know what it was saying. The words were loud and quick, spoken in fear and some anger. The voice was harsh. [3] The low voice said several times "Heaven help us!" and once "My God".

---

1. **party** : (here) group of people.
2. **foreigner** : someone from another country.
3. **harsh** : unpleasant, cruel, rough.

Jules Mignaud, banker, says that Madame L'Espanaye had some property. [1] She had an account with his bank. She made frequent deposits in small sums. Three days before her death she took out the sum of 4,000 francs. The bank paid her the sum in gold and sent a clerk to her house with the money.

Adolphe Le Bon, clerk to Mignaud and Son, says that at 12 noon he accompanied Madame L'Espanaye to her house with the 4,000 francs in two bags. He did not see anyone in the street at that time.

William Bird, tailor, is an Englishman. He has lived in Paris for two years. He was one of the first to go up the stairs. He heard the voices and also a sound, like the sound of people fighting. The shrill [2] voice was very loud. He believes it was German although he does not speak the language. Perhaps the voice of a woman.

Four of the above-named witnesses also said that the door of the room where they found the body of Mademoiselle L'Espanaye was locked from the inside. Everything was perfectly silent. When they opened the door there was nobody there. The windows of both

1. **property** : buildings or land.
2. **shrill** : high and unpleasant.

the back and front room were closed and locked from inside. A door between the two rooms was closed but not locked. The door from the front room into the corridor was locked with the key on the inside. A small room in the front of the house at the end of the corridor was open. This room was full of old beds and boxes. The police searched the whole house.

Some of the witnesses say that only three minutes passed between the time they heard the angry voices and the moment they forced the door of the room. Others think the interval [1] was as long as five minutes.

Alfonso Garcia, undertaker, [2] says that he lives in the Rue Morgue. He was one of the party who entered the house but he did not go upstairs. He was too afraid. He heard the voices arguing but he could not hear what they said. The low voice was that of a Frenchman. The high voice was an Englishman. He is sure of this although he does not understand English.

Alberto Montani, baker, says he was one of the first to go upstairs. He heard the voices clearly. The low voice was that of a Frenchman. He thinks that the shrill voice was speaking Russian. He has never spoken to anyone from Russia.

Several witnesses said that the chimneys of all the rooms on the fourth floor were too small for a human being to enter them. The

---

1. **interval** : a period between two events.
2. **undertaker** : person whose job is to deal with the bodies of people who are dead and to arrange funerals.

apartment had no back door for a killer to make his escape while the party were coming up the stairs. The body of Mademoiselle L'Espanaye was so firmly pushed up the chimney that it took four or five of the party to remove it.

Paul Dumas, doctor, says that he saw the bodies in the early morning. They were both lying in the room where the daughter was found. The young lady's body was covered in cuts and bruises. The throat was greatly marked. The face was discoloured and the tongue was partially bitten through. According to M. Dumas, the girl's death was the result of strangulation. The body of the mother was horribly mutilated. All the bones of the right arm were broken. The whole body was badly bruised and discoloured. It was not possible to say what the cause of these injuries was. Possibly a heavy wooden club [1] or an iron bar or a chair. Any large, heavy object could produce these results in the hands of a powerful man. But it would be impossible for a woman. The head of the old lady was separate from the body. Her throat was cut, probably with a razor.

This is the strangest murder case that Paris has ever seen. As usual, the police know nothing. But there is not one single clue to help them.

1. **club** : thick heavy stick.

The witnesses

1 In Chapter 2 you can find the testimonies of 11 people. Identify the characters and write down their names and professions.

a . Pauline Dubourg   laundress

b . ...

c . ...

d . ...

e . ...

f . ...

g . ...

h . ...

i . ...

j . ...

k . ...

What did the witnesses say?

2 Are these statements true (T) or false (F)? Correct the false ones.

	T	F

a . Pauline Dubourg said that although she had known the victims for three years, she still didn't know what Madame L'Espanaye's job was and she had never seen anybody in the house apart from the lady herself and her daughter.

b . Pierre Moreau said that he had known Madame L'Espanaye for seven years. He added that the victims had led a very extravagant [1] life despite their lack of means and had often had friends staying over at weekends.

1. **extravagant** : spending more than necessary.

c. Isidore Muset said that when he had gone to the
house, the gate had been locked and he had a hard
time trying to force it. He explained that he had heard
loud screams and two voices speaking in a foreign
language, possibly Arabic.

d. Henri Duval said that he himself had heard two
voices but he didn't agree with Isidore Muset as to
the language they were speaking. In fact he thought
that the high voice was that of an Italian and the other
voice was that of a Frenchman. He was sure that
neither voice belonged to the owners of the apartment.

e. Monsieur Odenheimer said that the screams had
lasted for about five minutes. He said that he had heard
the two voices too but couldn't work out what they
were saying since his French was not good enough to
understand their rapid speech.

f. Jules Mignaud stated that Madame L'Espanaye was not
very well-off. In fact, three days before her death she
had borrowed the sum of 4,000 francs.

g. Adolphe Le Bon declared that at twelve noon he had
accompanied Madame L'Espanaye to the bank to collect
the money. The street had been absolutely deserted
at the time.

Four more witnesses

3 **Summarise in your own words the remaining testimonies.**

h. William Bird said that ....

i. Alfonso Garcia stated that ....

j. Alberto Montani declared that ....

k. Paul Dumas explained that ....

Reported speech

4 So far you have read the testimonies of the official witnesses. However, a small newspaper published the statements of other neighbours who didn't want to be interviewed at first.
Turn the following statements into reported speech.

a. (Monsieur Brelle) "Madame L'Espanaye never said hello to me!" Monsieur Brelle said that...

b. (Madame La Fayette) "I've been living in this area for fifteen years but I have never met the lady you are referring to."

c. (Paul Leroux) "If I knew something I would tell you but, believe me, I didn't hear a single noise."

d. (Bettie Bellini) "I don't know what you are on about. If you don't leave me alone, I'll call the cops." [1]

e. (Claude Puselle) "I was sleeping when the horrible murders happened... I was having a nightmare actually."

f. (Madame Sorelle) "This is the most appalling [2] thing which has ever happened in this area. It used to be a respectable Quartier, you know!"

Writing a letter

5 You happened to be there when the whole thing happened. Obviously you didn't see the assassin [3] but you also heard the two voices and witnessed the horrible scene in Madame L'Espanaye's apartment.
Write a letter to a friend recounting [4] this terrible experience.

---

1. **cops** : police officers.
2. **appalling** : shocking.
3. **assassin** : killer.
4. **recounting** : describing.

Before you read

1 Listen to the first part of the dialogue between the narrator and Dupin in Chapter 3. Here you have a short report. Fill in the gaps with a suitable word.

The police have .... Adolphe Le Bon, the ....
from the bank, but Dupin doesn't trust them very much. In fact,
according to him, there is no .... in the way they work,
other than the method of the .... Sometimes they get
.... results but most of the time these are simply thanks to
.... and .... work. However, there are times when
these .... are not enough and their strategies ....

Vidocq for instance, is a professional policeman but he always makes
the same ...., concentrating too closely on an object and
therefore missing the object as a ....

In Dupin's opinion the .... is not always at the bottom of
the .... In fact, he believes the truth is often at the
.... of things. Will his intuition lead him to the solution of
the mystery?

Now read the text and check your answers.

CHAPTER 3

At the Scene of
the Crime

The story of the murders in the Rue Morgue continued in the evening edition of the newspaper. "It says here that the police have arrested and imprisoned Adolphe Le Bon, the clerk from the bank," I said.

"The Parisian police are clever, but no more than that," Dupin replied. "There is no method in the way they work, other than the method of the moment. The results they get are surprising but most of the time they are obtained simply thanks to diligence [1] and hard work. But when these qualities are not enough, their strategies [2] fail. Vidocq, for example, was a good policeman. But he always made the same mistake. His investigations were always too intense. [3] He

1. **diligence** : careful hard work.
2. **strategies** : detailed plans for achieving something.
3. **intense** : using a lot of effort.

couldn't see clearly because he held the object too close. Perhaps he saw one or two details clearly but in doing this he couldn't see the object as a whole. It is possible for an investigation to be too profound. [1] The truth is not always at the bottom of a well. [2] In fact, I believe that the truth is often at the *surface of things.*

END

"As for these murders," Dupin continued, "we will go and see the house with our own eyes. I know the Police Commissioner. [3] It will not be difficult to obtain permission."

Dupin obtained permission and we went immediately to the Rue Morgue. It was late in the afternoon when we arrived at the house. It was an ordinary Parisian house. Before we went inside we walked around the building. Dupin examined the whole area with great attention. Then we entered the house.

We went upstairs, to the room where they had found the body of Mademoiselle L'Espanaye. To my surprise, the bodies of the two women were still there. Dupin examined everything including the dead bodies. We then went into the other rooms, accompanied by a policeman. We stayed in the house until it began to get dark. Then we began the journey home. On the way home Dupin stopped for a moment to visit the offices of one of the daily newspapers.

My companion did not speak about the murders again until about noon the next day.

"Did you notice anything *peculiar* at the scene of the murders?" he asked me.

---

1. **profound** : deep, extreme.
2. **well** : hole in the ground from which you can obtain water.
3. **commissioner** : an important official.

"No, nothing *peculiar*," I said. "Only the things I read about in the newspaper."

"The newspapers know nothing!" he declared. "It seems to me that they consider this case insoluble [1] for the very reason that renders [2] it easy to solve. I mean the bizarre [3] character of the murders. The police cannot understand the fact that there is no obvious motive for the atrocity [4] of the murders. They are also confused by the angry voices because there was no one upstairs apart from the body of Mademoiselle L'Espanaye. And there was no way to leave the building apart from the stairs – the stairs that the investigating party of neighbours were going up. The disorder of the room; the incredible violence of the murders; the mutilation of the old lady's body. These things are incomprehensible [5] to the police. But the police have made a mistake. They think the case is difficult when in fact it is only very strange. In this case we must not ask 'What has happened?' We must ask 'What has happened *that has never happened before?*' "

I looked at Dupin in silent astonishment.

"I am waiting for a person to come here. I don't think he is responsible for these terrible murders but he is involved in them in some way. I look for the man here – in this room – every moment. Perhaps he will not arrive but the probability is that he will."

Dupin continued, looking at the wall as he spoke.

"We know that the witnesses heard two voices arguing and that these voices were not the voices of the victims. And if they

---

1. **insoluble** : difficult to solve.
2. **renders** : causes.
3. **bizarre** : very strange and unusual.
4. **atrocity** : cruelty.
5. **incomprehensible** : extremely difficult to understand.

are not the voices of the victims they must be – or one of them must be – the voice of the killer. Let us now consider the testimony. Did you notice anything *peculiar* about it?"

"Well, all the witnesses said the low voice was that of a Frenchman. But they all had different opinions about the other voice."

"Yes, that was the testimony they gave. But it was not the *peculiarity* of the testimony. The witnesses, as you say, agree about the identity of the low voice. But regarding the high voice, the strange thing is not the fact that they disagreed. The strange thing is that each of them, an Englishman, a Spaniard, an Italian, a Hollander [1] and a Frenchman thought it was *the voice of a foreigner*. They were all sure that it was not the voice of one of their countrymen. But each of these witnesses also says that the voice was speaking a language *that they do not know*.

1. **Hollander** : a native of Holland (usually Dutchman).

At the Scene of the Crime

The Frenchman says it was the voice of a Spaniard, *but he does not speak Spanish.* The Hollander *does not speak French* but says that the high voice was that of a Frenchman. The Englishman thinks that it was the voice of a German and *does not understand German.* The Spaniard is sure that it was the voice of an Englishman *but he does not know English.* The Italian believes that it was the voice of a Russian *but he has never spoken to anyone from Russia.*"

Dupin continued. "From these conflicting testimonies we can conclude that the voice was very strange. In addition, one witness says the voice is harsh rather than shrill. And two others say it is quick and unequal. No witness could distinguish [1] any words it said.

"I do not know," continued Dupin, "if this is easy for you to understand, but I tell you now that the part of the testimony regarding the two voices *is enough to enable us to find the solution to this mystery.*"

1. **distinguish** : identify.

Dupin's visit to the scene of the murders

1 The narrator and Dupin decided to go and see the scene of the murders with their own eyes and managed to obtain permission. Summarise their visit to Madame L'Espanaye's apartment and their journey back in seven steps.

a. They arrived at the house late in the afternoon.

b. ....

c. ....

d. ....

e. ....

f. ....

g. ....

Unfinished sentences – Dupin's analytical mind

2 Dupin thinks that the police are going about things the wrong way. Complete his opinions about their attitude towards the crimes.

a. The police think that ....

b. They cannot ....

c. They are puzzled ....

d. They don't see how ....

e. They find ....

f. They are wrong in thinking ....

g. They shouldn't be asking ....

Prediction

3 The enigma seems very difficult to solve yet Dupin appears to be quite confident. Try to predict the following mysteries.

- Who was the man that Dupin was waiting for?
- From the conflicting testimonies, we can understand that one of the two voices spoke a language which nobody seemed to be sure about. A strange language, harsh and shrill... What language was it?

  Remember that Dupin said:
  'The Part of the testimony regarding the two voices is enough to find the solution to this mystery!'

Past perfect

4 Late in the afternoon the narrator and Dupin arrived at Madame L'Espanaye's flat. What had happened at the apartment some hours before? Look back at Chapter 1 and use the verbs in brackets to recount the murders.

a. (break in) ....
....

b. (break) ....
....

c. (open, scatter about) ....
....

d. (bruise) ....
....

e. (strangle) ....
....

f. (mutilate) ....
....

Singular or plural?

5 Form sentences with the following words choosing between the singular and plural form.

a. Dupin's thesis [1] is / are ....

b. The police has / have ....

c. Dupin's family was / were ....

d. The evidence of the murders is / are ....

e. The furniture in Madame L'Espanaye's apartment was / were ....

f. The goods in the house wasn't / weren't ....

g. The people living in the area was / were ....

h. The hair found in the apartment was / were ....

6 Can you find six hidden words connected with the idea of Murder?

K	H	K	I	L	L	E	R	N	M
I	O	S	Q	Z	O	P	T	K	X
L	M	A	S	S	A	C	R	E	N
L	I	S	L	A	A	X	B	H	J
I	C	F	A	U	S	V	W	M	B
N	I	P	U	I	M	S	J	H	D
G	D	M	G	L	J	L	U	S	I
R	E	Z	H	S	C	G	H	T	N
Q	T	H	T	A	E	D	C	R	E
W	T	R	E	Q	C	H	A	Z	P
D	L	L	R	F	Y	O	U	S	Q

1. **thesis** : the main theory of a person.

The Art of the Detective Story

Poe has been unanimously [1] acclaimed as a master of the short story, a genius of analytical imagination, a virtuoso [2] of both logic and hallucination. [3] Although they may seem contradictory, it is the fusion of these elements that give Poe's work its unique flavour. The English writer D.H. Lawrence once compared Poe to a scientist, saying "he is reducing his own self as a scientist reduces a salt (...) It is an almost chemical analysis of the soul and consciousness.',
It is this synthesis [4] of art and science, narrative and mathematics that forms the basic recipe for all Poe's short stories. Nevertheless, these works belong to two quite separate narrative traditions, that of the grotesque [5] and that of ratiocination. [6]

Together with other short stories like *The Mystery of Marie Roget*, *The Gold Bug* and *The Purloined Letter*, *The Murders in the Rue Morgue* belongs to the tales of Ratiocination, as Poe himself defined them. From a modern perspective we could say that they are the forerunners of the detective story, measuring the detective's deductive skills against often bizarre mysteries and enigmas. *The Gold Bug*, for example, is based around the deciphering [7] of a cryptogram [8] – a piece of secret writing hidden in another text, while *The Purloined Letter*'s logic is in a way very similar to that

1. **unanimously**: unitedly.
2. **virtuoso** : a person who shows exceptional skill.
3. **hallucination** : illusion of seeing or hearing something.
4. **synthesis** : mixture.
5. **grotesque** : strange and unpleasant.
6. **ratiocination** : process of logical and methodical reasoning.
7. **deciphering** : interpreting.
8. **cryptogram** : message written in code.

which is to be found in *The Murders in the Rue Morgue*. Both feature the same protagonist, [1] the French detective C. Auguste

Arthur Conan Doyle.

Dupin, who reveals his method of investigation, that is to say his art of analysis, in a few words "Perhaps it is the very simplicity of the thing which puts you at fault", and thus by looking at the surface of things is able to solve a puzzle that was elementary yet beyond the police's hurried perception of crime as always being complicated. We can hear the echo of Dupin in Conan Doyle's Sherlock Holmes, for whom the business of deduction is best summed up in his famous catchphrase, "Elementary, my dear Watson."

The tradition of detective as logician which Poe inaugurated [2] still has its followers, even among contemporary detective story writers. But the somewhat artificial formal puzzles which writers such as Poe, Conan Doyle and Jacques Futrelle gave their heroes to solve – the so-called "locked room" mysteries – have more or less disappeared. Partly responsible for this

1. **protagonist** : the most important character in a story.
2. **inaugurated** : began.

was the development of the hardboiled [1] detective, the big city private investigator whose archetype [2] was Raymond Chandler's Philip Marlowe. The type of mystery the hardboiled detective is called to investigate is almost the inversion of the tradition of Poe. Here, what begins as a relatively simple matter of, say, finding a missing person soon becomes a terrifyingly complex and illogical web which the detective can barely understand. In his essay "The Art of Murder," Chandler acknowledged Dashiell Hammett as the true inventor of the hardboiled detective. Tired of the artificiality of the "locked-room" puzzles Hammett "gave murder back to people who committed it for a reason." But in doing so, paradoxically, [3] Hammett introduced the "irrational" into the art of the Detective story.

1 **Now answer the following questions.**

a. What are some of the elements which can be found in Poe's short stories?

b. What are the two categories we can group his texts into?

c. How was Dupin's approach to crime different from that of the police?

d. How did hardboiled detective fiction differ from the classical detective story?

e. In what way is hardboiled fiction an "inversion" of the Poe tradition?

f. Hammett "gave murder back to people who committed it for a reason." What do you think he meant by this?

1. **hardboiled** : someone who is tough and doesn't show much emotion.
2. **archetype** : a typical example of something.
3. **paradoxically** : in a contradictory way.

45

CHAPTER 4

The Mystery Unfolds

upin continued, "Let us now imagine that we are in the room in the Rue Morgue. What is the first thing we should look for? The way the murderers escaped. Madame L'Espanaye and her daughter were not killed by ghosts. So how did the murderers escape? Let us now examine the possible means of exit. One thing is clear. When the party came up the stairs, the assassins were either in the room where they found Mademoiselle L'Espanaye or the room next to it. So we have only two rooms to look for exits. The police have made a complete search of the apartment. But when we were there, I decided to make my own search.

"Both doors going from the rooms into the corridor were locked with the keys inside. Then there are the chimneys. As we

have already seen, these are too small to permit a person to climb up them. In fact they are too small to permit a cat to climb up them. So the only possible exit that remains are the windows. The murderers couldn't escape through the windows at the front of the house because somebody in the street would see them. Therefore they passed through the windows in the back room.

"There are two windows in this room. One of them is completely visible. [1] I examined this window and found that it was nailed shut. [2] It was impossible to raise it. Then I examined the other window. It too was nailed shut. The police concluded from this that the murderers did not use the windows to escape. Even if they did manage to raise the windows they could not replace the nail again from the outside. So the police did not even try to remove the nail from the second window. But I knew that the windows were the only means of exit.

"I went to the first window and with difficulty I took the nail out. Once the nail was out it was easy to raise the window. But the police were right. It was impossible to replace this nail from the outside. Then I turned to the other window. It looked exactly the same as the first. Once again I pulled the nail. But when I took it out, I saw that what I had in my hand was only a part of the nail. I raised the window and saw that the bottom part of the nail

1. **visible** : able to be seen.
2. **nailed shut** : closed securely with nails and so impossible to open.

remained in the window sill. [1] The nail was broken and I saw that the fracture [2] was very old. Now I understood everything. This window only *appeared* to be locked. And when the assassins closed it from the outside, the nail once again appeared secure. The police didn't see this because they didn't try the second window. They assumed *from its appearance* that it too was securely locked.

"The next question is how the murderers got down from the fourth floor to the ground. Before we went up to Madame L'Espanaye's apartment, I had a walk around the building. Less than two metres from the window I have just spoken about, there

1. **window sill** : a ledge or shelf along the bottom of a window.
2. **fracture** : crack.

is a lightning rod [1] which runs to the ground. From this rod it is impossible to reach the window itself. But then there are the shutters on the windows. These shutters are very interesting. They are in the form of a single door, but the top part is composed of horizontal wooden bars which provide excellent hand-holds. [2] Each shutter is about a metre wide, which means that if it is extended fully back to the wall, the shutter of our window is less than a metre from the lightning rod. It is possible that someone with great strength and agility [3] could jump from the lightning rod to the shutter. Then he could use his feet to push himself from the wall and close the shutter. And if the window was open he could even enter the room.

"But after all I have said, I want you to consider the *extraordinary agility* necessary to do this. It is possible but – and here is the important point – you would need to have an almost *animal-like* agility to do it.

"Now we must put together the two peculiar facts we have. The first is this act of animal-like agility that I have just spoken about. The second is the extremely peculiar voice that the witnesses spoke about, the voice that was both shrill and harsh, and also unequal. And of course you remember that none of the witnesses could agree about its country of origin."

At that point I felt that I almost understood what Dupin was saying but I could not make the final connection.

---

1. **lightning rod** : long, thin piece of metal which attracts lightning and allows it to reach the ground safely.
2. **hand-holds** : small spaces where you can put your hands.
3. **agility** : (physical) quickness.

"You see," Dupin continued, "that I have moved the focus of my investigation from how the assassins got out to how they got in. Because I believe they used the same route in both cases. But now let us consider the inside of the apartment. Here, the peculiar thing is why the assassins did not take the gold. Almost all the 4,000 francs that Monsieur Mignaud talked about were in bags on the floor. This money cannot be the motive. It is simply a coincidence that the money was delivered [1] to the house on the same day as the murders. Stranger coincidences than this happen every day of our lives. If the gold was the real motive, then the murderer must be a complete idiot. [2] It is impossible to believe that after murdering the two women, he forgot to take the gold.

"So now we have three points: the strange voice, the incredible agility and the absence of any motive for the murders."

1. **delivered** : taken.
2. **idiot** : a stupid person.

Find the right words

1 Here are some definitions. Find the words they refer to in the text.

a. A small group of people associated together in any occupation: ....

b. The act of examining closely or exploring with a view to finding something: ....

c. A small spike [1] of metal: ....

d. The part at the foot of a window: ....

e. A slender [2] bar of metal or another material: ....

f. A cover to close a window: ....

g. High-pitched and piercing: [3] ....

True or false?

2 Dupin's speculations [4] are beginning to lead us towards the solution of the crime. These are some of his brilliant intuitions but some of them are not accurate. Correct the false ones.

	T	F
a. The first thing to look for in order to find the solution was the way the murderers had escaped.	☐	☐
b. Both doors facing the corridor were locked from the outside.	☐	☐
c. The chimneys were quite wide.	☐	☐
d. The murderers, in order not to be seen, must have passed from the windows in the back room.	☐	☐
e. The three windows in the back room were nailed shut.	☐	☐
f. One of the windows turned out to be broken which meant that it only appeared to be locked.	☐	☐

1. **spike** : a narrow thin shape with a sharp point at one end.
2. **slender** : thin.
3. **piercing** : (describing a sound) high and unpleasant.
4. **speculations** : guesses.

g. Less than one metre from this window there was a ☐ ☐
lightning rod which ran to the ground.

h. The shutters could provide perfect hand-holds, as ☐ ☐
long as the murderers were very agile and strong.

i. The assassins were absolutely indifferent to [1] the 4,000 ☐ ☐
francs that were lying on Madame L'Espanaye's bed.

Three conclusions

3 Putting together the peculiar facts that he had discovered, Dupin
came to three conclusions.
What were they?

1st ....

2nd ....

3rd ....

Very analytical... but what does he mean?

4 The narrator is amazed by how quickly Dupin has arrived at a solution
to the crime but there are some things he doesn't quite understand.
Write five queries starting with the following question words.

a. Who ....?

b. Where ....?

c. What ....?

d. How ....?

e. Why ....?

1. **absolutely indifferent to** : not caring about at all.

52

Relative clauses

5 Link the two sentences with a relative pronoun. Omit it when you can!

a. Dupin was a brilliant investigator. He didn't take long to find out what had really happened.

....

b. The murderer had almost decapitated [1] Madame L'Espanaye. His strength must have been remarkable.

....

c. Dupin searched the flat. He found a lot of interesting things there.

....

d. Dupin had a lot of intuitions. He told them to the narrator.

....

e. The narrator was astonished by his reconstruction of the events. He was not as insightful as Dupin.

....

f. The police inspected the windows. They appeared to be locked.

....

g. Each window had shutters. Their width was about one metre.

....

h. The murderer must have been of an extraordinary agility. He climbed up the wall.

....

i. One of the two voices was extremely shrill. Dupin talked about it.

....

1. **decapitated** : cut off the head of a person.

Before you read

1 Listen to Dupin's speech in Chapter 5 and fill in the gaps with the missing words.

"But let us look now at the murders .... Here is a woman .... to death and then .... up into a chimney. Would an ordinary murderer push his victim up a chimney? Don't you think that it is a bit .... ? A little absurd even? Do you think that it is something a normal human being would do?

"Remember also that it .... four men to .... her down from the chimney. The murderer .... be incredibly strong. And there is other .... to suggest this almost superhuman .... On the fireplace there were several lengths – very thick lengths – of human .... Now you and I know that it is .... difficult to .... even twenty or thirty hairs together from the human head. But our murderer pulled perhaps .... a million hairs from the head of Mademoiselle L'Espanaye."

Now read the text and check your answers.

54

BLACK CAT ENGLISH CLUB
The Commercial Press (Hong Kong) Ltd.
9/F, Eastern Central Plaza,
3 Yiu Hing Road, Shau Kei Wan,
Hong Kong

BLACK CAT ENGLISH CLUB

Membership Application Form

BLACK CAT ENGLISH CLUB is for those who love English reading and seek for better English to share and learn with fun together.

Benefits offered:
- *Membership Card*
- *Book discount coupon*
- *English learning e-forum*
- *English learning activities*
- *Black Cat English Reward Scheme*
- *Surprise gift and more...*

Simply fill out the application form below and fax it back to 2565 1113 or send it back to the address at the back.

Join Now! It's FREE exclusively for readers who have purchased *Black Cat English Readers* !

(Please fill out the form with **BLOCK LETTERS**.)

The title of Black Cat English Reader/book set that you have purchased: _____

English Name: _____ (Surname) _____ (Given Name)

Chinese Name: _____

Address:

Tel: _____ Fax: _____

Email: _____

Sex: ❏ Male ❏ Female (Login password for e-forum will be sent to this email address.)

Education Background: ❏ Primary 1-3 ❏ Primary 4-6 ❏ Junior Secondary Education (F1-3)
❏ Senior Secondary Education (F4-5) ❏ Matriculation
❏ College ❏ University or above

Age: ❏ 6 - 9 ❏ 10 - 12 ❏ 13 - 15 ❏ 16 - 18 ❏ 19 - 24 ❏ 25 - 34
❏ 35 - 44 ❏ 45 - 54 ❏ 55 or above

Occupation: ❏ Student ❏ Teacher ❏ White Collar ❏ Blue Collar
❏ Professional ❏ Manager ❏ Business Owner ❏ Housewife
❏ Others (please specify: _____)

As a member, what would you like **BLACK CAT ENGLISH CLUB** to offer:

❏ Member gathering/ party ❏ English class with native teacher ❏ English competition
❏ Newsletter ❏ Online sharing ❏ Book fair
❏ Book discount ❏ Others (please specify: _____)

Other suggestions to **BLACK CAT ENGLISH CLUB**: _____

Please sign here: _____ (Date: _____)

Visit us at Quality English Learning Online http://publish.commercialpress.com.hk/qel

CHAPTER 5

"Caught"

 ut let us look now at the murders themselves. Here is a woman strangled to death and then pushed up into a chimney. Would an ordinary murderer push his victim up a chimney? Don't you think that it is a bit excessive? [1] A little absurd [2] even? Do you think that it is something a normal human being would do?

"Remember also that it took four men to pull her down from the chimney. The murderer must be incredibly strong. And there is other evidence to suggest this almost superhuman strength. On the fireplace there were several lengths – very thick lengths – of human hair. Now you and I know that it is extremely difficult to pull even twenty or thirty hairs together from the human head. But our murderer pulled perhaps half a million hairs from the head of Mademoiselle L'Espanaye. And then the throat of the old lady was not simply cut: the head was completely separated from

END

---

1. **excessive** : more than expected.   2. **absurd** : unreasonable.

the body. But the instrument the murderer used was a simple razor. Then we must also consider the terrible ferocity [1] of these actions.

"Now we almost have a complete picture of the murderer: incredibly agile with superhuman strength, brutally [2] ferocious but without motive, inhuman in his reasoning and actions and with an extremely strange voice that is foreign to the ears of men from many different countries. What is your opinion?"

"A madman," I said. "Some maniac [3] escaped from a psychiatric hospital." [4]

"An interesting idea," said Dupin, "but madmen come from some nation. It is true their language is often incoherent [5] but it does contain recognisable words. And the hair of madmen is not like the hair I now hold in my hand. I found this piece of hair in Madame L'Espanaye's hand. What do you think?"

"Dupin!" I said, completely shocked; "this is no *human* hair."

"I didn't say that it was," said Dupin, "but before we decide its true nature I want you to look at this drawing. It is a *fac-simile* [6] drawing of the bruises and fingermarks that were found on the throat of Mademoiselle L'Espanaye.

"You see that this drawing gives us the idea that the hands of the murderer easily encircled the throat. They did not move until she was dead. Now, take this cylinder of wood which is approximately the diameter of the throat and wrap the drawing around it."

I followed Dupin's instructions.

"Now I want you to try to place your fingers over the fingermarks of the murderer."

1. **ferocity** : violence.
2. **brutally** : cruelly and violently.
3. **maniac** : a mad person.
4. **psychiatric hospital** : a place where mentally-ill people stay.
5. **incoherent** : not clear or logical.
6. **fac-simile** : an exact copy.

"Caught"

I tried to do this but it was impossible. The hands on the paper were too big, the fingers too long.

"This," I said, "is the mark of no *human* hand."

"Now read this article that I found in a book on ethology." [1]

I took the book that Dupin offered me and read. It was a detailed description of the large orang-outang [2] of the East Indian islands. I knew very well the beast's enormous size, its incredible strength and ferocity and its capacity [3] to imitate [4] human sounds and actions. I now understood completely the horrors of the murders in the Rue Morgue.

"The description of the fingers is exactly the same as those in the drawing," I said. "The orang-outang is the only animal that could make these marks. Also the hair that you showed me is identical to that of the beast described in the book. But I cannot understand the details of the mystery. For example, the fact that there were *two* voices arguing and one of them was clearly that of a Frenchman."

"This is true," said Dupin, "and we also know that it was the voice of the Frenchman who said the words 'My God!'. Certainly this Frenchman knew about the murders. It is probable, however,

1. **ethology** : the science of animal behaviour.
2. **orang-outang** : large ape with long arms.
3. **capacity** : ability.
4. **imitate** : copy the behaviour of someone.

that he was innocent [1] of the bloody act itself. I imagine that the orang-outang escaped from him and that he followed it to the house in the Rue Morgue. But after the confusion of the incident I do not believe that he recaptured [2] it. It is still free. But this is just my intuition. If the Frenchman is innocent of the murders, the advertisement which I left at the office of *Le Monde* last night will bring him to our house."

He gave me a newspaper and I read:

CAUGHT

In the Bois du Boulogne early in the morning of —— (the morning after the murder) a very large red orang-outang. The owner of the animal (believed to be a sailor from a Maltese ship) may come to collect it at No. – Rue – , Faubourg St. Germain.

"How could you possibly know," I asked, "that the man was a sailor, and from a Maltese ship?"

"I do *not* know it," said Dupin. "I am not *sure* of it. But I found this piece of ribbon at the bottom of the lightning rod. It is the type of ribbon that sailors use to tie their hair in a ponytail. If you look at the knot [3] you will see that it is the kind that only sailors

---

1. **innocent** : not guilty.
2. **recaptured** : captured again.
3. **knot** :

can tie. And this particular knot is peculiar to the Maltese.

"Our man is innocent of the murders," Dupin continued, "but he knows about them. Certainly he will hesitate [1] before replying to the advertisement – and before coming here and asking for the orang-outang. But he will also think: 'I am poor; my orang-outang is of great value – particularly to someone poor like me; why should I be afraid? Why should I lose it because of some imagined danger? After all, the beast was found in the Bois du Boulogne – a very great distance from the scene of the murders. And no-one could possibly think that an animal was responsible for them. Above all, the advertiser knows me. I don't know how much he knows about the atrocity, but if I do not collect the animal I will attract suspicion, both to it and to me. And I do not want to do that. Therefore I will answer the advertisement and get the orang-outang back and keep it with me until everyone has forgotten this horrible incident.' "

1. **hesitate** : pause (before doing something).

What happened in Chapter Five?

1 **Answer the following questions.**

a. What was the narrator's opinion regarding the identity of the murderer?

....

b. Why didn't Dupin find it convincing?

....

c. What did Dupin get the narrator to do with the drawing?

....

d. What was their conclusion after doing so?

....

e. What did a book on ethology have to do with the murders in the Rue Morgue?

....

f. Whose were the two voices?

....

g. What was the *Le Monde* advertisement about?

....

h. Who wrote it and why?

....

i. What did the piece of ribbon prove?

....

j. From whom was Dupin expecting a visit soon?

....

A bizarre advertisement

2 **Imagine you have found a small fox in your garden. Write an advertisement to be published in the local newspaper to find the owner of the animal.**

A C T I V I T I E S

Understanding the mind of the sailor

3 Dupin was very insightful and he knew that the sailor would go to the place mentioned in the ad. What led him to believe so?

a . ...
....

b . ...
....

c . ...
....

d . ...
....

e . ...
....

Role-play

4 When Dupin went to the office of *Le Monde*, he found a very curious [1] and inquisitive [2] secretary...

You are Dupin and your friend is the secretary who is puzzled by the advertisement and asks a lot of questions about it. Remember that Dupin doesn't mind answering questions because he is really proud of his discoveries.

1. **curious** : eager to know or learn.
2. **inquisitive** : asking too many questions about something or someone.

61

Apes Going Ape

The king of all apes who go ape [1] is of course King Kong. The original King Kong featured in the 1933 classic film of the same name. The beast, who is captured on a hunting expedition, [2] is brought to New York to be displayed for all the world to see. Needless to say, Kong is not too happy about having been torn from his homeland and shipped to America for the entertainment of the masses [3] and soon evades [4] his captors, [5] kidnapping the hunter's girlfriend and taking her to the top of the Empire State Building. The final sequence which features the wounded Kong swiping [6] at

*King Kong and Fay Wray, by Merian C. Cooper
and Ernest B. Schoedsak, 1933.*

1. **go ape** : (slang) become crazy.
2. **expedition** : a long and carefully organised journey.
3. **the masses** : all the ordinary people in society.
4. **evades** : escapes from someone.
5. **captors** : people who capture animals or people.
6. **swiping** : hitting.

the aeroplanes that buzz around him, while holding the screaming heroine [1] in one of his huge fists is surely one of the most famous in the history of cinema. Surrounded by the bad acting of the cast, Kong, brilliantly animated by Willis O' Brien, is easily the most interesting character in the film and emerges as a tragic hero for modern times.

A slightly more eloquent [2] ape is the narrator of Franz Kafka's short story, "A Report for an Academy" who, imprisoned in a cramped cage, decides that his only "way out" is to evolve, which he does by imitating people. Reporting the story to his judges – the Academy – he remarks ironically: "your own apehood, gentlemen, in so far as you have anything of the sort behind you, cannot be farther removed from you than mine from me."

In a clever reversal of the idea of the ape as our savage [3] ancestor, the 1960s science fiction film *Planet of the Apes*, the first and best

Planet of the Apes by Franklin J. Schaffner, 1967.

1. **heroine** : the main female character in a film.
2. **eloquent** : giving a strong message.
3. **savage** : extremely wild.

of a series of adaptations of Pierre Boulle's pedantic [1] novels, has the apes ruling over what appears to be a simian [2] version of the Roman empire, while human slaves provide amusement and opportunities for scientific study.

Somewhat more realistically perhaps, chimpanzees [3] have provided the villains for two recent thrillers: In *Link*, (1985) a psychopathic [4] cigar-smoking chimp chases a terrified woman around an isolated mansion after his master mysteriously disappears, while George Romero's infinitely superior *Monkey Shines*, (1988) tells the story of Ella, a chimpanzee who has been injected with human brain tissue to make her more intelligent. But with her new semi-human intelligence Ella also learns human emotions such as rage, frustration, jealousy and hate, all of which make her final descent [5] into madness and resulting razor-wielding [6] rampage [7] much more believable and thus more terrifying.

1 Are these sentences true (T) or false (F)? Correct the false ones.

	T	F
a. "Going ape" means *going out of control.*	☐	☐
b. King Kong was brought to New York for scientific study.	☐	☐
c. In Kafka's story the ape decides to become like a human being.	☐	☐
d. Pierre Boulle's novels are set at the time of the Roman Empire.	☐	☐
e. *Monkey Shine* is a thriller.	☐	☐
f. Romero's film is unrealistic.	☐	☐

Before you read

1 **Listen carefully!**

Listen to Chapter 6 and answer the following questions.

a. How old was the orang-outang?

....

b. Where did the story begin?

....

c. What was the orang-outang holding in its hand when it ran away from the sailor's place?

....

d. What happened to Adolphe Le Bon?

....

e. What happened to the orang-outang?

....

1. **pedantic** : giving too much attention to rules and small details.
2. **simian** : monkey-like or ape-like.
3. **chimpanzees** : African apes.
4. **psychopathic** : mentally ill and dangerous.
5. **descent** : a gradual change towards behaviour that is very bad.
6. **razor-wielding** : holding a razor.
7. **rampage** : violent and destructive behaviour.

CHAPTER 6

A Sailor's Story

A t that moment we heard the sound of footsteps on the stairs. But now the visitor seemed to hesitate and we heard him going back down. Dupin was moving to the door when we heard him coming back up. This time he did not hesitate but came to our door and knocked.

"Come in," said Dupin in a friendly voice.

A man entered. He was a sailor, a tall, muscular man.

"Good evening," he said.

"Sit down, my friend," said Dupin. "I imagine you are here for the orang-outang. A remarkable beast. How old is he?"

"I don't know exactly," the sailor replied, "maybe four or five. Have you got him here?"

"Oh no," said Dupin. "We cannot keep him here. He is at a

stable [1] in the Rue Dubourg. You can get him in the morning."

"Thank you," said the sailor. "I will of course pay you a reward for finding him."

"That's very kind of you," said Dupin. "Now let me see. What shall I ask for this service? Ah yes, I know. My reward shall be this. You will tell me everything you know about these murders in the Rue Morgue."

As he said these words, Dupin walked slowly to the door and locked it, and put the key in his pocket. He then took a pistol from his pocket and placed it calmly on the table.

The sailor fell back into his chair, trembling with fear. His face went as white as death itself. He did not say a word.

"My friend," said Dupin in a kind voice. "There is no need to be alarmed. I know that you are innocent of the murders in the Rue Morgue. But you must admit that you are involved in them in some way. You have nothing to hide. But every principle of honour obliges [2] you to confess [3] all that you know. An innocent man is now in prison, charged with these murders. You must tell us who the real perpetrator [4] is."

"I will tell you all I know about this business, but you will not believe me. I do not really believe it myself. But I am innocent."

1. **stable** : building in which horses are kept.
2. **obliges** : forces.
3. **confess** : admit.
4. **perpetrator** : the person responsible for a crime.

A Sailor's Story

The sailor was silent for a moment. Then he began to tell the story.

"It began on the island of Borneo. Our ship stopped there on the way to India. A friend of mine captured an orang-outang but he fell sick and died. I became the beast's master. I took it with me on board the ship, where it stayed until we got back to Paris. Then I hid it in my apartment. I knew that it was valuable so I decided to sell it.

"The next evening I went out with some sailor-friends. When I got back in the early morning I could not believe what I saw. There, sitting in my chair, was the orang-outang. Its face was covered in shaving foam [1] and in its hand was my razor. It sat there looking at itself in the mirror. I realized that the beast was trying to shave. But all I could think about was the razor in its hand. I was terrified. For a minute I didn't know what to do. Then I took the whip [2] that I used to control the orang-outang. When he saw it he was afraid, but before I could do anything he ran out of the

---

1. **shaving foam** : a substance that men put on their faces before they shave.

2. **whip** :

69

room and down the stairs, and then jumped out of an open window into the street.

"I followed the orang-outang down street after street. The city was deserted. It was three o'clock in the morning and everyone was in bed. Finally I trapped it in an alley [1] at the rear [2] of the Rue Morgue. But then the beast saw a light shining from the open window of Madame L'Espanaye's room on the fourth floor of her house. Running to the building, it saw the lightning rod, which it climbed with unbelievable agility. Then, with one long arm, it grasped [3] the shutter which was against the wall and used it to swing itself into the room.

"At first I was happy. The beast was trapped and it would be easy to capture it now. But I was afraid for the occupants of the house. So I decided to follow it. With some difficulty, I climbed up the lightning rod, but when I got to the level of the window I could not reach the shutter. I could only look into the room. I was horrified by what I saw and

1. **alley** : narrow street.
2. **rear** : back part.
3. **grasped** : took hold of.

heard. The two women were screaming. I saw the orang-outang take Madame L'Espanaye by the hair and cut her throat with the razor. The movement of its arm was so powerful that the woman's head was almost separated from her body. On seeing her blood, it became furious and jumped upon the girl, putting its hands around her throat until she too was dead. Just then, it saw me through the window and was suddenly afraid.

"It became agitated [1] and started throwing the furniture around the room. I could see it wanted to hide the evidence of its terrible acts, so it took the body of the girl and pushed it up the chimney. Then it turned to the body of the old woman which it threw out of the window.

"As the orang-outang approached [2] the window I was terrified. I climbed back down the lightning rod and ran home as fast as I could. I wanted to get as far away from the beast as possible."

"So the words that the party on the stairs heard were your exclamations of horror," I said.

"And the shrill, harsh voice was that of the beast," Dupin added.

And that is almost the end of the story. Dupin and I went to the police and told them everything. They immediately released Adolphe Le Bon, the clerk from the bank. And as for the sailor – well, he finally caught his orang-outang and sold it to the city zoo for a very large sum of money.

1. **agitated** : anxious and not calm.
2. **approached** : came near.

True or false?

1 Are these sentences true (T) or false (F)? Correct the false ones.

	T	F
a. The sailor was a skinny man.	☐	☐
b. Dupin expected the sailor to pay a large sum of money as a reward.	☐	☐
c. Dupin put a pistol on the table and this scared the sailor.	☐	☐
d. Dupin knew the sailor was innocent.	☐	☐
e. The orang-outang was captured in Borneo by the sailor himself.	☐	☐
f. When the sailor arrived from meeting his friend, the beast was trying to shave.	☐	☐
g. When the beast entered the house in Rue Morgue, the sailor was unable to follow it.	☐	☐
h. One of the two voices was the sailor's.	☐	☐
i. The sailor was really shocked by the beast's actions.	☐	☐

The key to the mystery

2 The editorial board of *Le Monde* have convinced Dupin to write a short article about the murders entitled:

THE TRUTH ABOUT THE RUE MORGUE!!!

Write a short but detailed article about what really happened.

Third conditionals

3 **What would/wouldn't have happened if...**
Make sentences using the following cues.

a. the sailor / not go / Borneo
b. the orang-outang / not find / the razor
c. the window / properly locked
d. Madame L'Espanaye / more fit
e. the orang-outang / not escape from the house

Speaking – role-play

4 **Work with your friend. You are a journalist and your friend is the sailor. Act out the interview. Remember to include as many details as possible.**

The story

5 a. Did you like the story?
b. Why/Why not?
c. If you had to describe it with three adjectives, how would you define it?

The characters

6 **Try to describe each character in a few sentences.**

The Narrator ....
Dupin ....
The sailor ....
The orang-outang ....

Paris in the 1800's

Napoleon Bonaparte, crowned emperor in 1804, had great projects for the capital. He decided to improve the River Seine. The Pont des Arts, made entirely of metal, had just been finished. During his reign three kilometres of quays [1] were built from the Louvre to the Tuileries, and around the Cité. Rue de Rivoli with its arches was created.

The Emperor then undertook the construction of the Arc de Triomphe which was completed many years later in 1835, and the church La Madeleine, in neoclassic style.

With the increase of the population new markets and numerous slaughter-houses [2] were created to meet the needs of the Parisians.

After the fall of Napoleon, work projects slowed down as the kings

A quay of the River Seine in 1894.

1. **quays** : long platforms beside the river.
2. **slaughter-houses** : places where animals are killed for their meat.

who followed did not have Napoleon I's ambitious projects and the coffers [1] had been left empty.

Slowly, however, the coffers filled up again and new living areas were created inside the capital and outside its walls.

Thanks to new water supply systems hygiene improved greatly. Numerous public baths were created, and they were used by the middle classes, who didn't have bathrooms at home. Men and women went to the public baths on alternate days. Doctors suggested taking baths, but not more than once a month!

Gas street lighting was introduced and people were able to stroll [2] in the streets until very late at night.

Gas street lamps.

Horse-drawn public carriages, 1835.

1. **coffers** : money that the government has to spend.
2. **stroll** : walk in a slow and relaxed way.

Cafés and theatres opened. In 1828 the first comfortable public
carriages appeared which guaranteed a regular transport service. The
mail was picked up several times a day from more than 200 mailboxes.
The demand for food in Paris was constantly increasing. In spite of the
cholera [1] epidemic [2] of 1832, the population continued to grow during
the reign of Louis-Philippe. A great number of nobles who had fled
during the Revolution returned.

The capital had its own food supply thanks to the dairies, cultivated
land and vineyards inside and outside the city walls.

Markets, food and wine shops, mediocre and luxurious cafés, and
prestigious restaurants grew everywhere in the capital.

There were some strange constructions along the Seine: flat-bottomed
boats with a basin in the middle where washerwomen washed clothes
and often hung them to dry. These were the laundry boats of the
washerwomen of the Seine.

Small trades developed and artisans flourished. The "bouquinistes", or
used-books merchants, invaded the quays and bordering area of the
Seine, as exasperated bookshop owners looked on. Small industries,
which produced luxury products such as perfumes and lace, began to
appear.

The development of the railroad promoted commerce and banking, and
industrial fortunes expanded.

Along with this economic growth, poverty, alcoholism and contagious
diseases proliferated at various social levels. Social work grew slowly
and important hospital projects were implemented.

Teaching developed too, thanks to the initiative of the church, but
illiteracy [3] remained high. Children were sent to work in factories

1. **cholera** : a serious infection of the bowels.
2. **epidemic** : (disease) spreading quickly among many people in the same
   place for a time.
3. **illiteracy** : state of being unable to read and write.

instead of attending school. Louis-Philippe ordered the embellishment [1] of
Place de la Concorde. The obelisk [2] of Luxor (Egypt) was erected [3] in 1836.
In 1855 and 1867, during the reign of Napoleon III (Second Empire),
two universal Expos took place. Parisians and visitors from all over the
world admired the great industrial and technical discoveries of the 19th
century.

Napoleon III asked the prefect Haussmann to design and restructure the
city with wide streets, and demolish the old, unhealthy areas.

Huge "boulevards" [4] were created that made it easy to move about in
the capital.

Big markets were rebuilt with new materials: iron and cast iron.

The prestigious Théâtre de l'Opéra was erected in this period.

Théâtre de l'Opéra.

1. **embellishment** : making something more beautiful.
2. **obelisk** : a tall stone column.
3. **erected** : built.
4. **boulevards** : wide roads in a city.

A new kind of shop was born: the department store, such as the "Bon Marché" and the "Belle Jardinière". People were very enthusiastic because of two great novelties: [1] the entrance was free and the prices were indicated on the products.

In 1870 the Franco-Prussian War caused the fall of the Second Empire. A disastrous peace treaty for France followed.

The new Assembly of Deputies chose to meet in Versailles and not in Paris. This awkward move caused discontent mainly among the poor classes who had already suffered greatly in the war. A General Council met at the Hotel de Ville and the "Insurrection [2] of the Comune" took place. The rebels and the deputies confronted each other in Paris, which was in flames. The rebels were shot. After a difficult beginning, the new Republic rebuilt a big part of the capital, and tried to make the Parisians and the entire world forget the victory of the Prussians, the insurrection of the rebels, and the civil war.

Le Louvre department store.

The last great work projects which Haussman had begun during the Second Empire were finally completed. Examples of these were the department stores "La Samaritaine" and "Le Louvre". The steam

---

1. **novelties** : qualities of being new and unusual.
2. **insurrection** : rebellion.

engine greatly improved public transport: trams, funicular railways [1] (there are many hills in Paris), buses and boats transported numerous travellers who moved rapidly from one part of the city to the other.

In 1873 an enormous votive [2] church was built on the hill of Montmartre. It was made of a special kind of marble that turned white under the rain. It was the basilica of "Sacré Coeur", built to thank Jesus for having saved Paris during the war against the Prussians. A strange tower slowly rose in the skies of Paris. Passers-by were worried about it. Engineer Eiffel, the designer, promised that the tower would be ready for the Expo of 1889 to commemorate [3] the centennial [4] of the French Revolution. Eiffel also reassured the people that the tower would be taken down right after the Expo!

The Eiffel Tower.

The nineteenth century was drawing to a close. Students and factory workers went to dance in the popular dance halls located at Montmartre and other suburbs of Paris.

At the "Moulin Rouge" one could admire the dancers of the can-can, who danced to the lively rhythms of Offenbach's music.

This was the beginning of the "Belle Epoque", a long period of light-heartedness and joy, that started in the 1880's and was brutally interrupted by the first World War.

---

1. **funicular railways** : a type of railway that travels up and down steep slopes.
2. **votive** : given or done because of a promise made to God.
3. **commemorate** : do something to show that one remembers an important event in the past.
4. **centennial** : the 100th anniversary.

1 Answer the following questions.

a. What was built during Napoleon Bonaparte's reign?

b. Why did work projects slow down after the fall of Napoleon?

c. How often did doctors suggest that people took baths?

d. What were the "bouquinistes"?

e. What new kind of shop was born?

f. What kinds of public transport were there?

g. What was special about the marble of the basilica of "Sacré Cœur"?

h. What was the Eiffel Tower built for?

i. How long did the "Belle Époque" last?

Boulevard Saint-Michel.

The
Purloined
Letter

CHAPTER 1

The Search for the Letter

I t was just after dark one windy evening in the autumn of 18__. I was with my friend C. Auguste Dupin, in the library of his Paris apartment. For one hour at least we had been sitting in profound silence. Dupin seemed content just to watch the smoke rising from his pipe. As for myself, I was thinking about the affair of the Rue Morgue. More particularly I was thinking about how quickly Dupin had solved the case when the police could do nothing. It was therefore something of a coincidence when our old acquaintance [1] Monsieur G___ of the Paris police arrived at the door.

G. explained that he had called to consult us, or rather to ask the opinion of my friend, about some official business that had caused a great deal of trouble.

"And what is the difficulty now?" I asked. "No more murders, I hope?"

"Oh no, nothing like that. The fact is, the business is very simple indeed, and I am sure we can manage it ourselves. But I thought Dupin would like to hear the details of the case, because it is so very odd." [2]

"Simple and odd," said Dupin.

"Yes, in a way, though not exactly. The fact is, we have all been extremely puzzled [3] because the affair is so simple and yet… we cannot understand it."

"Perhaps it is the very simplicity of the case which you do not understand," said my friend.

"What nonsense you talk sometimes, Monsieur Dupin," said the Prefect, laughing.

"Perhaps the mystery is a little *too* simple," Dupin continued.

"Oh good heavens! What an absurd idea!"

"A little *too* obvious."

"Ha! ha! ha!" laughed our visitor, deeply amused by Dupin's suggestion.

"And what after all is this business all about?" I asked.

"I will tell you in a few words," said the Prefect of police, "but

1. **acquaintance** : someone you know slightly, but not well.
2. **odd** : strange.
3. **puzzled** : confused.

before I begin I must warn you that this affair demands the greatest secrecy. If anybody knew that I was telling you about it, I would probably lose my job."

"Proceed," [1] I said.

"Well then; I have received personal information from a very high official that a certain document of the utmost [2] importance has been purloined [3] from the royal apartments. The individual who took it is known; he was seen to take it. It is also known that he still has the document."

"How do you know this?" asked Dupin.

"We know, because of the nature of the document, and because if it had passed out of the robber's possession it would have certain... ahem... 'dramatic' consequences. Since these consequences have not yet arrived, we can assume that the robber still has the document."

"Can't you be a little more explicit?" [4] I said.

The policeman hesitated for a minute.

"If this document were revealed to a third person – an exalted [5] personage [6] whom I cannot name – it would cause grave damage to the honour and reputation of its owner – who is herself an important royal personage. This fact gives the robber power over her. He can use the document as a form of blackmail." [7]

"But this power would depend on the robber knowing that the owner of the letter knew he had taken it."

1. **proceed** : continue.
2. **utmost** : greatest.
3. **purloined** : stolen.
4. **explicit** : clear and exact.
5. **exalted** : very important.
6. **personage** : an important person.
7. **blackmail** : when you make a person pay you money to keep his secrets.

The Search for the Letter

"The thief," said G. "is the Minister D____, an unscrupulous [1] individual. The method of the theft was ingenious. [2] The document – it is a letter to be honest – had been received by the royal personage while she was alone in the royal bedroom. As she was reading it, she was suddenly interrupted by the entrance of her husband, the other personage I mentioned before, the person that she wanted to conceal [3] it from. She tried to throw the letter into a drawer, without success, and so she was forced to lay it, open as it was, on a table. Fortunately only the side with the address on it was visible and so the letter was not noticed. It was then that the Minister D____ entered. His sharp eye immediately perceived [4] the letter and recognised the address. When he saw the anxious state of its owner, he knew at once her secret. After some routine business transactions, he produced from his pocket a letter similar to the one lying on the table and opened it, pretending to read the contents. He then casually placed his letter next to the one on the table and continued conversing with the lady and her husband about political matters. Then just before leaving the room he took the lady's letter from the table. Of course she saw him take it but she could say nothing. Her husband was standing next to her, and naturally she did not want

---

1. **unscrupulous** : behaving in an unfair or dishonest way.

2. **ingenious** : clever and skilful.

3. **conceal** : hide.

4. **perceived** : saw.

him to know about the letter. And so the minister left the room
with the vital [1] document in his pocket, leaving his own letter –
which was of no importance – upon the table."

"So it is true," said Dupin. "The robber knows that his victim
saw him take the letter. He has her in his power."

"Yes," said the Prefect of Police. "And for the last few months
he has been using that power for political purposes, to a very
dangerous extent. The lady who was robbed is now convinced that
she must get the letter back. But this cannot be done openly. That
is why she has put me in
charge of the case."

1. **vital** : extremely
   important.

·D·

The Search for the Letter

"Quite right," said Dupin, through the smoke. "I cannot imagine that the Paris police has a better agent."

"You flatter [1] me," said the Prefect, "but perhaps there is some truth in what you say."

"It is clear," said I, "that the minister still has the letter in his possession. If he used it it would no longer give him any power."

"True," said G. "and that is why I decided to conduct a meticulous [2] search of the minister's house. The problem was how to search without his knowledge. It would be extremely dangerous

1. **flatter** : praise someone even though you do not mean it.
2. **meticulous** : very careful and detailed.

if he realized we were looking for the letter. Fortunately for us the minister is often absent from home all night, and he does not have many servants. The servants' bedroom is quite far from the minister's rooms, and the servants are usually drunk at night.

"As you know, I have keys which can open any room or cabinet in Paris. So every night for the last three months we have been searching D____'s house. But after all our efforts I must now conclude that the minister is a more astute [1] individual than I am. I can assure you that we have investigated every corner of the premises, every little space where it would be possible to conceal the letter, but still we cannot find it."

"But is it not possible," I said, "that he has hidden it somewhere else, somewhere other than his own house?"

"No, I don't think that's very likely," said Dupin. "He needs to have the document near to him. It must be instantly available so that he can produce it whenever it becomes necessary to do so."

"If that is the case, then the letter must be somewhere in the minister's house. For it is not possible that it is on his person."

"Absolutely not," said the Prefect of Police. "My men have stopped and searched him twice under my own supervision."

"Obviously the man is not a complete fool," said Dupin. "He surely knew that he would be searched."

"Not a complete fool, no," said G. "But he is a poet, which I consider to be only one grade higher than a fool."

"Tell us the details of your search," I said.

---

1. **astute** : clever and quick to see how to take advantage of a situation.

"Well, we took our time and we searched everywhere. I have a lot of experience in these matters. We divided the building into its separate rooms and for each of these rooms we spent an entire week searching. First we examined the furniture. We opened every possible drawer. To a well-trained police officer the existence of a secret drawer is impossible: there is only a certain amount of space in each cabinet. After the cabinets we examined the chairs. We probed [1] the cushions with long needles. Then we removed the tops from the tables."

"Why did you do that?"

"Sometimes a person who wishes to conceal a document removes the top of a table and excavates [2] one of the legs. He then pushes the document into the cavity [3] and replaces the top. The bottoms and tops of bedposts [4] are used in the same way."

"But you could not possibly remove – you could not dismantle [5] all the pieces of furniture in which it would be possible to hide a letter. A letter can be rolled into a thin spiral tube, no thicker than a knitting needle. [6] In this form it could be inserted into the rung [7] of a chair for example. You did not dismantle all the chairs?"

1. **probed** : (here) examined.
2. **excavates** : removes.
3. **cavity** : an empty space.
4. **bedposts** : vertical supports at the corners of a bed.
5. **dismantle** : take apart.
6. **knitting needle** :
7. **rung** :

The Purloined Letter

"Certainly not; but we did better – we examined the rungs of every chair in the house, and indeed the jointings [1] of every other piece of furniture, with the help of a very powerful microscope. If there had been any changes in the jointings, any disorder in the glueing, we would have noticed it immediately."

"I presume you looked at the mirrors, between the frames and the glass plates, and also the beds and bedclothes as well as the curtains and carpets?"

"Of course; and when we had finished examining the furniture we searched the house itself. We divided its surface into numbered compartments; [2] then we scrutinized [3] each individual square centimetre of the house, and also of the two adjoining houses." [4]

"The two adjoining houses!" I exclaimed. "You must have had a great deal of trouble."

"Yes we had, but the reward for finding the letter is very high."

"You included the grounds around the houses?"

"All the grounds are paved [5] with brick. It was easy to search them. We simply had to check that the grass between the bricks was undisturbed."

"Of course you looked through D___'s papers and examined his books."

1. **jointings** : places where two parts of something meet.
2. **compartments** : separate parts.
3. **scrutinized** : examined something very carefully.
4. **adjoining houses** : the houses on either side.
5. **paved** : covered.

"Certainly; we opened every file and turned over every page of every book. We also measured the thickness of each book cover and examined the bindings [1] with the microscope. If any of the bindings had been altered [2] in any way we would have noticed."

"You explored the floors beneath the carpets."

"Yes. We removed every carpet and examined the floorboards with the microscope."

"And the wallpaper?"

"Yes."

"You looked in the cellars?"

"Yes we did."

"Then," I said, "you have made a miscalculation and the letter is not in the house."

"I think you are right," said the Prefect. "And now Dupin, what do you think I should do?"

"I think you should search the house again."

"That would be absolutely pointless," [3] said the Prefect. "I tell you, as sure as I live and breathe, the letter is not in the house."

"I have no better advice to give you," said Dupin. "I presume you have an accurate description of the letter."

"Oh yes!" said the Prefect. He took a notebook out of his pocket and began to read a detailed account of the internal, and especially of the external appearance of the stolen document. Soon afterwards, he thanked us for our time and left the apartment. He looked more depressed than I had ever seen him before.

---

1. **bindings** : the parts of books where the pages are attached to the front cover.
2. **altered** : changed.
3. **pointless** : useless.

Monsieur Dupin

1 The hero of the story, which takes place some time after the case of the Rue Morgue, is again Monsieur Dupin. How is he presented at the beginning of the story? Use your own words to complete the following sentences.

a. It was a windy ....

b. The narrator and Dupin were ....

c. For one hour ....

d. Dupin was watching ....

e. The narrator was thinking ....

f. Suddenly ....

Second or third conditional?

2 This time the case concerns a letter which has been purloined. The whole affair is surrounded by an air of secrecy. Complete the following sentences with the right conditional.

a. If anybody knew that the policeman Monsieur G. told Dupin about the case, ...

b. ...., the police wouldn't have been so puzzled.

c. If the document had passed out of the robber's possession, ...

d. ...., certainly the case wouldn't have been solved.

e. If the Minister D. realized the police were looking for the letter, ...

f. If the minister hadn't often been absent from home, ...

What type of letter?

3 The object of the investigation is a letter but the narrator is not very explicit about what type of letter it is. Try to imagine three types of letter. Remember that the content must be important enough to upset the political situation.

a . ....

b . ....

c . ....

Where is the letter?

4 The police have searched the house thoroughly but couldn't find the precious ¹ document. Make predictions about where the letter can be.

a . It must be ....

b . It can be ....

c . It could be ....

d . It may be ....

e . Perhaps ....

f . It is probably ....

g . In my opinion ....

h . I think it could ....

1. **precious** : of great value.

Searching step by step

5 The search of the minister's house was very meticulous. Are the following sentences true (T) or false (F)? If they are false, correct them with the right piece of information.

		T	F
a.	Monsieur G. was an expert in conducting searches.	☐	☐
b.	First of all they divided the building into its separate rooms.	☐	☐
c.	For each of these rooms they spent two days searching.	☐	☐
d.	They examined everything except the chairs.	☐	☐
e.	They removed the tops from the tables.	☐	☐
f.	They used a microscope to search his desk.	☐	☐
g.	They looked between the frames and the glass plates of the mirrors.	☐	☐
h.	They looked under the carpets and behind the curtains.	☐	☐
i.	They examined the main building and the cellar but didn't examine the two adjoining houses.	☐	☐
j.	They measured the thickness of each book cover as well as their bindings.	☐	☐

Role-play

6 Work with your friend. You are Dupin and your friend is the policeman. Dupin is very inquisitive about what the police did when they searched the minister's house and the policeman is very precise in answering the questions.

CHAPTER 2

Dupin's Techniques

A bout a month later, the Prefect paid us another visit. He found us occupied in the same manner as before. He sat down and began talking about the general business of the week.

After a few minutes I interrupted him:

"Well, but G, what about the Purloined Letter?"

The Prefect's face turned pale.

"I searched the house again, as Dupin suggested, but it was a waste of time, as I knew it would be."

"How much was the reward offered, did you say?" asked Dupin.

"A very large reward – a very *liberal* [1] reward – I don't like to say how much; but I will say one thing, that I would give my own cheque for fifty thousand francs to anyone who could give me that letter. The fact is, the matter is becoming more and more urgent every day; the reward has recently been doubled. But even if it

1. **liberal** : giving in a generous way.

were trebled [1] I could do no more than I have done already."

"Do you think so?" said Dupin. "I think, G___ , that there is some more that you could do to find the letter, eh?"

"How? – in what way?"

"You might, for example, take advice."

"My dear sir, I would happily pay for such advice. I would really give my fifty thousand francs to anyone who could help me."

"In that case," replied Dupin, "you can write me a cheque now. When you have signed it, I will give you the letter."

I was astonished. The Prefect appeared absolutely incredulous. [2] For some minutes he remained silent and motionless, looking at my friend with open mouth and eyes that seemed to jump out from their sockets. [3] Then he suddenly took a pen and after some hesitation signed a cheque for fifty thousand francs and gave it to Dupin. Dupin quickly examined the cheque and put it in his wallet. Then he opened a drawer in his desk and took a letter from it which he gave to the Prefect. The Prefect looked at the letter. His face was a perfect agony of joy as he opened it and quickly scanned its contents. Without saying another word, he ran out of the apartment still holding the letter in his hand.

When he had gone, Dupin began to explain.

"The Parisian police," he said, "are very capable in their way. They are persevering [4] and ingenious and they know their job well. When G_____ recounted to us his mode of searching D_____'s house, I was sure he had made a satisfactory investigation in as far as he was able."

---

1. **trebled** : increased three times in amount.
2. **incredulous** : not willing to believe.
3. **sockets** : two hollow bony parts where your eyeballs are.
4. **persevering** : showing continued steady effort to achieve an aim.

"In as far as he was able?" I said.

"Yes. The techniques he used were not only the best of their kind, they were also executed to absolute perfection. If the letter had been hidden within the limits of G_____'s search, he would certainly have discovered it."

I laughed, but Dupin seemed quite serious in all he was saying.

"The techniques," he continued, "were the best of their kind and were well executed. The problem was that *they were not suitable for this case.* They could not be applied to this particular thief. The Prefect has a set of extremely ingenious resources and he thinks he can use them to solve every single case in the same way. But he continually makes the mistake of being either too deep or too superficial, for the case in question; there are many schoolboys who are better thinkers than he is. I knew one about eight years of age who had great success in the game of 'even or odd'. [1] This game is simple and is played with marbles. [2] One player holds in his hand a number of marbles and

---

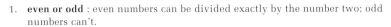

1. **even or odd** : even numbers can be divided exactly by the number two; odd numbers can't.
2. **marbles** : small balls of coloured glass.

asks the other player whether the number is even or odd. If his guess is right, the guesser wins one: if he is wrong he loses one. The boy I refer to won all the marbles of the school. Of course he had a principle which he used to make his guesses. He observed and measured the intelligence of his opponents. [1] For example, let's imagine his opponent is a complete idiot. The idiot holds up his closed hand and asks the boy 'are they even or odd?' Our schoolboy replies 'odd' and loses; but the next time he wins. How does he do this? He simply says to himself, 'The idiot had them even the first time and his amount of intelligence is just sufficient to make him change them to odd the second time.' So he guesses odd and wins. Now with an idiot a degree more intelligent than the first he would reason like this: 'This boy sees that the first time I guess odd. Now the second time he will immediately propose a simple variation from even to odd, like the first idiot did.' But then a second thought will tell him that this variation is too simple, and finally he will decide to keep the marbles even as before. 'I will therefore guess even.' – he guesses even and wins. And so on. Now this method of reasoning which the other boys call 'luck' – what is it exactly?"

"It is simply a question of the boy trying to think like his opponent," I said.

"Precisely," said Dupin. "The boy identifies himself with his opponent's intellect." [2]

"And the success of this identification depends on how accurately he can measure the intellect of his opponent?"

"Yes, for its practical value it depends on this," replied Dupin. "This is where the Prefect and his men make the mistake. They do

1. **opponents** : competitors.
2. **intellect** : cleverness.

not identify themselves with the intellect of their opponent. They do not even try to measure his intellect. They consider only *their own* ideas of ingenuity; and in searching for anything hidden, they consider only the ways in which *they* would have hidden it. In part they are right to do so, for their ingenuity is an accurate representation of that of *the mass*; but when the intelligence of the thief is greater than their own – and often when it is lower, their investigations fail. They have no variation in principle in their methods. The most they ever do is to *extend* or *exaggerate* [1] these old methods, without considering the principles on which they are based. Let's look at the case of D_____. What did G____ do to vary the principle of action? What is all this drilling [2] and probing and sounding and scrutinizing with the microscope and dividing the building into numbered square centimetres? It is simply the exaggeration of the *application* of one set of principles, which are based on the Prefect's own ideas about human ingenuity. He thinks that all men will hide a letter in the same way, not necessarily inside the leg of a chair, but in some place which is equally difficult to reach. But such hiding places are imagined only by ordinary intellects. In all cases of concealment, it is presumed that the article will be concealed in a place which is hard to find. Thus its discovery depends not upon the intelligence of the seekers but upon their patience, perseverance and determination. You will now understand what I meant when I said that the Prefect would surely have found the letter if it had been placed within the limits of his examination – in other words, if the principle of its concealment was the same as the principle of the Prefect's investigation. But the Prefect has

1. **exaggerate** : expand.
2. **drilling** : regular activity.

been mystified, [1] and the reason for his defeat is that he believes the minister to be a fool, simply because he is a poet. The Prefect thinks: all fools are poets, but that does not mean that all poets are necessarily fools!"

"I thought it was the minister's brother who was the poet. The minister is a mathematician, isn't he?"

"No, he is both a mathematician and a poet. If he were just a mathematician he would not be able to reason well, and the Prefect would have found the letter."

"You surprise me when you say this. Mathematical reason has for centuries been regarded as the greatest form of reason."

"That is what mathematicians have made us believe, but it is an error."

"Explain."

"Mathematical axioms [2] are not axioms of general truth, but only of relation. And something which may be true for relation – of form and quantity for example – is often completely false in terms of morality, or of human nature in general. To put it simply, in human nature the sum of the parts is rarely equal to the whole. In life, there is always something more, some unpredictable element that cannot be reduced to an equation."

"Whatever do you mean?" I said, laughing. By now I was completely confused.

"I mean to say," said Dupin, "that if the minister had been only a mathematician, the Prefect would have been able to keep

---

1. **mystified** : confused, puzzled.
2. **axioms** : statements or ideas which people accept as being true.

the fifty thousand francs for himself. He would certainly have found the letter. But I knew the minister was also a poet. And so I tried to think like him, considering the circumstances he was in.

"Go on," I said.

"Remember that this man is a clever politician. He obviously knew the normal police procedures. He knew he would be searched and I'm sure he also knew that the police would search his house. The Prefect said that he was fortunate that the minister was often absent from the house at night. But I think that the minister was deliberately absent. He wanted the Prefect to  search the house. It was all part of his plan. He wanted to make the police think that the letter was not in the house. So he let them search it from top to bottom. Finding nothing, the police would be convinced that the letter was not there. The minister must have known how the police conducted their searches. He must have known that the police would examine all the normal hiding places, all the places the Prefect told us about. So obviously he could not hide the letter in any of these places. I realized that because of this he would be forced to choose *simplicity* as the principle for hiding the letter. You remember how the Prefect laughed when I suggested that the problem of the mystery was perhaps the fact that it was *too* obvious."

"Yes," said I, "I remember it well."

"There is a game," Dupin said, "which is played on a map. One of the players tells another to find a given word on the map –

The Purloined Letter

it could be the name of a town or a river or a monument [1] – any word that appears on the map. Now a beginner immediately seeks [2] the most difficult names with the most complicated spelling; but the expert selects those simple words which appear in huge letters and which stretch from one end of the map to the other. These, like street signs, are written in big letters, but they escape our attention, precisely because they are so excessively obvious. We simply don't see them."

"But what is the connection between this and the minister?"

"What is true of the physical world is also true of the moral world. This is something the Prefect cannot understand. He never once considered that the minister might have placed the letter *right under his very nose*, that he had perhaps hidden it in a *very visible place,* where nobody would think of looking for it. Or that perhaps the best way to hide the letter was *not to hide it at all.*"

1. **monument** : a statue or building built to remind people of a famous person or event.
2. **seeks** : looks for.

104

What happened in Chapter Two?

1 **Answer the following questions.**

a. What were Dupin and the narrator doing when the policeman
went back a month later?
....

b. How much did the policeman offer as a reward for the person
who could give him the letter?
....

c. What was the policeman's reaction when Dupin gave him the
letter?
....

d. What was wrong with the techniques used by the Prefect and his
men?
....

e. What were the minister's occupations apart from politics?
....

f. How could you sum up the main principle behind Dupin's logic?
....

Games

2 It is obvious that Dupin likes games, especially those which deal
with the art of logic and the intellect. In this part of the story he
talks about how a clever boy always managed to win the game
called 'even or odd'. Have you ever played it?
Let us try to sum up the rules of the game and the method to win.

a. This game is played with ....

b. One player holds a number of marbles and asks the other player
....

c. If the other player is right, he ....

d. On the other hand, if he is wrong, ...
....

e. In order to win you must first of all observe and measure ....
....

f. If you think the opponent is a complete idiot, and if you lose the
first time answering "odd", you have to ....

g. This is because the opponent will think that ....
....

h. On the other hand, if your opponent seems to be slightly more
intelligent than the first, you should ....
....

i. This is because this "idiot a degree more intelligent than the
first" will probably think that ....

Passive

3 **Turn the following sentences into the passive form. Try to judge when
it is appropriate to put the agent and when it is not necessary.**

a. The narrator asked the Prefect a question about the purloined
letter.
....

b. When the policeman went to see Dupin, they had searched the
house several times.
....

c. The police are offering a very large reward for finding the letter.
....

d. To his surprise, Dupin returned the letter to the policeman.
....

e. It surprised the Prefect that Dupin had found the letter.
....

f. The Parisian police had used the wrong techniques to solve the
crime.
....

Are you a patient person?

4 See if you can discover these jumbled [1] words taken from the first two parts of the story.

a. dpbteso

b. rmbaels

c. nceltaoemnc

d. nimtaeahmicta

e. zupzedl

f. wrader

g. rbreob

h. earevsepcner

i. rhesca

j. tnshkcise

Reward

5 Write a short note to be published in the local newspaper about the reward and the nature of the 'search'.

SUBSTANTIAL [2] REWARD!

A very important letter ....

....

....

....

....

....

1. **jumbled** : mixed together without any order.
2. **substantial** : large in amount.

Before you read

1 Use the words in the box to fill in the gaps in the text. Then read Chapter 3 to check your answers.

> head   minister   business   purloined   spectacles
>
> table   mantlepiece   books   compartments   letter
>
> tear   female   appearance   Prefect   seal

"With these ideas in my .... I went to visit the .... at his house. I wore a pair of special green .... to aid my vision. As the minister told me about the political .... of the day, I scrutinized his large writing .... on which lay several papers and letters, as well as one or two .... But there was no trace of the .... letter itself.

"As my eyes circled around the room I noticed a card-rack hanging from the centre of the .... In this rack, which had five or six ...., were a few visiting cards and a solitary letter. The letter was very soiled and crumpled and it was torn at the edge. It seemed as if someone had decided to .... it up but had then changed their mind. I looked closer to see what was written on it. It was addressed to the minister himself in .... handwriting. The letter had been thrown carelessly into one of the compartments.

"I knew immediately that this was the .... I was looking for. Certainly its .... was completely different from the description of the stolen letter that the .... had given us. The wax seal of this letter was large and black with D's insignia inscribed on it, whereas the purloined letter had a small red .... bearing the coat of arms of the Royal Family."

CHAPTER 3

The Solution of the Case

With these ideas in my head I went to visit the minister at his house. I wore a pair of special green spectacles [1] to aid my vision. As the minister told me about the political business of the day, I scrutinized his large writing table on which lay several papers and letters, as well as one or two books. But there was no trace [2] of the purloined letter itself.

"As my eyes circled around the room I noticed a card-rack [3]

1. **spectacles** : glasses.
2. **trace** : a sign that something was present.
3. **card-rack** :

109

The Purloined Letter

hanging from the centre of the mantlepiece. [1] In this rack, which had five or six compartments, were a few visiting cards and a solitary letter. The letter was very soiled [2] and crumpled [3] and it was torn at the edge. It seemed as if someone had decided to tear it up but had then changed their mind. I looked closer to see what was written on it. It was addressed to the minister himself in female handwriting. The letter had been thrown carelessly into one of the compartments.

1. **mantlepiece** : shelf over the fireplace.
2. **soiled** : dirty.
3. **crumpled** : full of folds and creases.

The Solution of the Case

"I knew immediately that this was the letter I was looking for. Certainly its appearance was completely different from the description of the stolen letter that the Prefect had given us. The wax seal of this letter was large and black with D's insignia [1] inscribed [2] on it, whereas the purloined letter had a small red seal bearing the coat of arms of the Royal Family. This letter was addressed to D_____ in a small female hand. The writing on the purloined letter was large and bold [3] and it was addressed to a lady. But it seemed

1. **insignia** : symbols of rank or authority.
2. **inscribed** : written.
3. **bold** : noticeable.

to me that these differences were excessive. The fact *that this letter was almost the opposite* of the letter the Prefect described made me suspicious. Then there was the fact that this letter was so clearly visible, in full view of every visitor. So visible, in fact, that nobody would see it, just like in the game I told you about."

"So what did you do?"

"I memorized the appearance of the letter and left the minister's house immediately, leaving a gold tobacco box on the writing table.

"The next morning I called at the minister's house again, saying that I had forgotten my tobacco. The minister seemed happy to see me again and we continued the conversation we had had the previous day. Suddenly we heard a loud noise coming from outside, the sound of a pistol shot followed by several screams. D_____ ran to the window to see what was happening. In the meantime I went over to the card rack, took the letter, put it in my pocket and replaced it with a *facsimile.*

"The disturbance in the street had been caused by a man with a gun. He had fired it above the heads of the crowd and everybody thought he was simply a lunatic [1] or a drunk. In fact I had employed him to carry out this diversion [2] to distract [3] D_____ and give me time to get the letter."

1. **lunatic** : a person who is mentally ill.
2. **diversion** : something that takes your attention away.
3. **distract** : stop somebody concentrating on something.

"But why didn't you simply take it on your first visit?"

"D____," replied Dupin, "is a desperate and dangerous man. If I had taken the letter openly he would probably have had me killed before I could leave the building. But there was another reason. For eighteen months the minister has had the owner of the letter in his power. Now the situation is reversed. [1] She has him in her power."

"How is that?"

"Because he still thinks he has the letter, thanks to my facsimile. He will proceed with his blackmail and his political career will be destroyed. It's ironic in a way too."

"In what way?"

"It was the minister who gave me the idea of the facsimile, when he stole the original letter. To get the letter back I just did the same thing he did when he stole it. You could even say that between our letters there is a perfect, ahem... correspondence."

"Very funny."

1. **reversed** : the situation is changed to the opposite of what it was before.

What happened in Chapter Three?

1 Are these sentences true (T) or false (F)? Correct the false ones with the right information.

	T	F
a. Dupin thought the minister was perfectly aware of the fact that the police would search his house.	☐	☐
b. The minister wanted the police to think that the letter was hidden somewhere in the house.	☐	☐
c. The Prefect's mistake was mainly in not understanding that the minister could have placed the letter in an absolutely obvious place.	☐	☐
d. Dupin went to see the minister three times.	☐	☐
e. Dupin pretended that he had forgotten his hat at the minister's house.	☐	☐
f. The minister realized that the letter had disappeared.	☐	☐

Dupin's strategic visit to the minister: step by step

2 Unlike the police, Dupin managed to spot the letter and take it with him. Try to describe each step using the following key words.

a. went

b. was wearing

c. scrutinized

d. noticed

e. addressed

f. knew immediately

g. left a tobacco box

h. went back

i. heard

j. replaced

The two letters

3 The letter Dupin finds is very different from the one the Prefect had described. What are the main differences?

Letter A	Letter B

Letter writing

4 In Chapter 1 you predicted three possible types of purloined letter. Now choose one and write it out. Remember the content has to be somehow very dangerous or crucial [1] for the royal personage who doesn't want her husband to read it.

1. **crucial** : extremely important.

An interview with Monsieur Dupin

5 Imagine you are Dupin. Answer the following questions that a very curious journalist asks you about the Purloined Letter.

JOURNALIST: Monsieur Dupin, you have done it again. How did you manage to solve another apparently unsolvable case?

DUPIN: ....

JOURNALIST: Okay, but let's start from the beginning. Who stole the letter and why?

DUPIN: ....

JOURNALIST: Mmm, I see. But didn't the police search his house?

DUPIN: ....

JOURNALIST: Were they careful in making the search?

DUPIN: ....

JOURNALIST: So what did you do?

DUPIN: ....

JOURNALIST: A facsimile? Why?

DUPIN: ....

JOURNALIST: Very clever! But how come the Minister didn't realize what you were up to?

DUPIN: ....

JOURNALIST: It really sounds like you, Monsieur Dupin! Just one more thing. How would you sum up your brilliant methods in a few words for our readers?

DUPIN: ....

EXIT TEST

Focus on the context

1 **Answer the following questions.**

a. When and where was Edgar Allan Poe born?

b. What were the names of his foster-parents?

c. Why was he disappointed when *Tamerlane and Other Poems by a Bostonian* was published?

d. What do some critics consider *The Murders in the Rue Morgue* to be?

e. When did Poe die?

Focus on the story

1 **For questions 1-12, choose the correct answer A, B, C or D.**

The Murders in the Rue Morgue

1. Where was Madame L'Espanaye's body found?

    A In the chimney.

    B On the bed.

    C On the floor.

    D In the small garden at the back of the house.

2. What did Alberto Montani claim?

    A That one of the voices was French and the other was English.

    B That one of the voices was French and the other was Russian.

    C That one of the voices was French and the other was Spanish.

    D That one of the voices was French and the other was Italian.

3. According to Dupin, how did the murderer get in?

    A By climbing up the stairs.

    B By entering through a door from a room into the corridor.

    C By climbing in through a window.

    D By climbing up the chimney.

4. What made Dupin come to the conclusion that the murderer was an orang-outang?

    A The size of the hands and the type of hair.

    B The type of hair and the size of the feet.

    C The number of bruises on Mademoiselle L'Espanaye's throat.

    D He found some footprints in the garden.

5. What made Dupin think that the owner of the orang-outang was a sailor?

    A He found a rope with a sailor's knot in it.

    B He found a sailor's map in the garden.

    C He found a ribbon with a sailor's knot in it.

    D He found a piece of paper with the name of a Maltese ship on it by the lightning rod.

6. What happened to the orang-outang at the end of the story?

    A It was taken back to the island of Borneo.

    B It was kept by the sailor.

    C It was arrested by the police.

    D It was sold to the city zoo.

The Purloined Letter

7. How did the Minister D____ manage to get hold of the letter?

    A He stole it in the middle of the night.

    B He put his own letter on the table, then took the other letter just before leaving.

    C He asked the important lady to give it to him.

    D He persuaded one of the servants to get it for him.

8. What special object did the police use to help them search for the letter?

    A A magnifying glass.

    B A lantern.

    C Special glasses.

    D A microscope.

9. How much was the original reward for finding the letter?

    A Ten thousand francs.

    B Twenty thousand francs.

    C Fifty thousand francs.

    D A hundred thousand francs.

10. Why did the Prefect give Dupin fifty thousand francs?

    A Dupin had given him the purloined letter.

    B Dupin had given him extra advice on where to look for the letter.

    C Dupin promised to deliver the letter that evening.

    D Dupin flattered and complimented the Parisian police for their excellent work.

11. What did Dupin say was used in the game of 'even or odd'?

    A Skittles.

    B Small stones.

    C Plastic balls.

    D Marbles.

12. Where did Dupin finally find the stolen letter?

    A On the Minister's large writing table.

    B In one of the Minister's books.

    C Behind the clock.

    D In the card-rack.

2 Read the summary of *The Murders in the Rue Morgue* and think of the word which best fits each space. Use only one word in each space. There is an example at the beginning (0).

Example: (0)   about

Dupin and the narrator read **(0)** .... the murders in the Rue Morgue in a newspaper. Mademoiselle L'Espanaye had been **(1)** .... and her body had been placed in the chimney. Her mother was found in the garden with her **(2)** .... cut. There appeared to be no motive for the murders.

A number of **(3)** .... gave their testimonies. They all agreed that they had heard a French voice but each witness claimed to have heard a second voice in a different foreign language, although none of them could agree on which one.

Dupin started his own investigation and realised that the **(4)** .... had got in and out of the house through a **(5)** .... in the back room. He realised that the hand **(6)** .... had strangled Mademoiselle L'Espanaye was enormous and **(7)** .... could not belong to a human being and **(8)** .... could the hair which had been found in Madame L'Espanaye's hand. This led him to believe that the murderer had been an orang-outang. By writing an **(9)** .... in *Le Monde,* he found the owner of the orang-outang, **(10)** .... recounted the whole story. The sailor himself was innocent and when he finally caught his orang-outang, he sold it to the city zoo.

The Murders in the Rue Morgue and The Purloined Letter

KEY TO THE ACTIVITIES AND EXIT TEST

SOME INFORMATION ABOUT EDGAR ALLAN POE

Page 12 – exercise 1

a. F – He was never officially adopted.
b. T
c. T
d. F – Poe began his literary career with a collection of poems.
e. F – Poe wrote a poem called "The Raven".
f. F – Poe's wife was only thirteen when they got married.
g. F – Poe's wife died of tuberculosis.
h. T

THE MURDERS IN THE RUE MORGUE

Page 19 – exercise 1

Possible answers:

a. Although Dupin came from a noble family, he was not very rich.
b. Books were his only passion and luxury.
c. Because of Dupin's financial difficulties, the narrator offered to pay the rent of the house they would share in Paris.
d. Darkness was one of Dupin's obsessions.
e. During the day they closed all the shutters to simulate the conditions of darkness.
f. At night they walked around the city looking for things to stimulate their imagination.
g. Thanks to his amazing analytical ability, Dupin could see directly into men's hearts and minds.

Page 19 – exercise 2

a. The narrator and Dupin were looking for the same book in a library when they met.
b. They read about the terrible events that had occurred in Quartier St. Roch in a newspaper.
c. The first people to know of the murders were the neighbours who had heard terrible screams coming from the house.
d. Among other things, the police found a razor covered in blood and bloody lengths of human hair.
e. The dead body of Madame L'Espanaye's daughter was found in the chimney.
f. She had been strangled.
g. The corpse of the old lady was in the back garden.
h. Her throat was cut and her body was mutilated.
i. The police didn't know what to think about these two horrible murders.

Page 20 – exercise 3

Open answers.

Page 21 – exercise 4

was born/ was acclaimed / found / abandoned / didn't prove / had already acted / included / didn't have / led / was struggling / acting / ended / was taken / stayed / had once been / died / was still living / was adopted / was sheltered

Page 22 – exercise 5

a. Since Dupin was not particularly well-off, the rent was being paid by the narrator.
b. If the voices hadn't been heard by the neighbours, the murders wouldn't have been found out.
c. As the group ran up the stairs, two angry voices were heard coming from the upper part of the house.
d. A horrible scene was discovered by the police in Madame L'Espanaye's flat.
e. The girl's body had been found in the chimney.
f. A clue that could help to solve the mystery hasn't been found yet.

Page 23 – exercise 1

laundress / victims / daughter / affectionate / paid / job / washing / servant / fourth / tobacconist / born / moved / quiet / believed / entered / porter / doctor

Page 30 – exercise 1

a. Pauline Dubourg – laundress
b. Pierre Moreau – tobacconist
c. Isidore Muset – policeman
d. Henri Duval – neighbour
e. Monsieur Odenheimer – restaurant owner
f. Jules Mignaud – banker
g. Adolphe Le Bon – bank clerk
h. William Bird – tailor

i. Alfonso Garcia – undertaker
j. Alberto Montani – baker
k. Paul Dumas – doctor

Page 30 – exercise 2

a. T
b. F – Pierre Moreau said that he had known Madame L'Espanaye for four years. He added that the victims had led a very quiet life despite their wealth and they never had visitors.
c. F – Isidore Muset said that when he had gone to the house, he had easily managed to open the gates with a piece of metal. He explained that he had heard loud screams and two voices speaking in a foreign language, one French and the other possibly Spanish.
d. T
e. F – Monsieur Odenheimer said that the screams had lasted for about ten minutes. He said that he had heard the two voices too and that one of them had said several times "Heaven help us!" and once "My God!".
f. F – Jules Mignaud stated that Madame L'Espanaye was quite well-off and that three days before her death she had taken out the sum of 4,000 francs.
g. F – Adolphe Le Bon stated that at 12 noon he accompanied Madame L'Espanaye to her house with the money in two bags.

Page 31 – exercise 3

h. William Bird said that he had heard two voices. One of them was perhaps the voice of a woman and was shrill, very loud and spoke in German.
i. Alfonso Garcia stated that he had heard the two voices arguing but could not hear what they said. One of them was speaking in French

while the other was speaking in English.

j. Alberto Montani declared that he had heard two voices. The low voice was that of a Frenchman while the other was speaking Russian.

k. Paul Dumas explained that the girl's death had been the result of strangulation and that the body of the mother had been horribly mutilated.

Page 32 – exercise 4

a. Monsieur Brelle said that Madame L'Espanaye had never said hello to him.

b. Madame La Fayette said that she had been living in the area for fifteen years but she had never met the lady in question.

c. Paul Leroux said that if he had known something, he would have told the journalist but he hadn't heard a single noise.

d. Bettie Bellini said that she didn't know anything at all. She told the journalist that if he didn't leave her alone, she would call the cops.

e. Claude Puselle said that she had been sleeping when the horrible murders had happened... and she had been having a nightmare.

f. Madame Sorelle said that it was the most appalling thing that had ever happened in the area. It had used to be a respectable Quartier.

Page 32 – exercise 5

Open answers.

Page 33 – exercise 1

arrested / clerk / method / moment / surprising / diligence / hard / qualities / fail / mistake / whole / truth / well / surface

Page 40 – exercise 1

a. They arrived at the house late in the afternoon.

b. They walked around the building.

c. They entered the house.

d. They went upstairs, into the room where the body of Mademoiselle L'Espanaye had been found.

e. Dupin examined everything.

f. They went into the other rooms.

g. They stayed in the house until it started to get dark. Then they went home.

Page 40 – exercise 2

Open answers.

Page 41 – exercise 3

Open answers.

Page 41 – exercise 4

a. Someone had broken in.

b. The furniture had been broken.

c. The drawers had been opened and the papers had been scattered about.

d. The daughter's dead body had been badly bruised.

e. The bruises on her throat suggested that she had been strangled.

f. The body of Madame L'Espanaye had been completely mutilated.

Page 42 – exercise 5

a. Dupin's thesis is ... (open answer).

b. The police have ... (open answer).

c. Dupin's family was/were ... (open answer).

d. The evidence of the murders is ... (open answer).

e. The furniture in Madame L'Espanaye's apartment was ... (open answer).

f. The goods in the house weren't ... (open answer).

g. The people living in the area were ... (open answer).

h. The hair found in the apartment was ... (open answer).

Page 42 – exercise 6

```
K  H (K  I  L  L  E  R) N  M
I  O  S  Q  Z  O  P  T  K  X
L (M  A (S) S  A  C  R  E) N
L  I  S  L  A  A  X  B  H  J
I  C  F  A  U  S  V  W  M  B
N  I  P  U  I  M  S  J  H  D
G) D  M  G  L  J  L  U  S  I
R (E) Z  H  S  C  G  H  T  N
Q  T (H  T  A  E  D) C  R  E
W  T  R  E  Q  C  H  A  Z  P
D  L  L (R) F  Y  O  U  S  Q
```

THE ART OF
THE DETECTIVE STORY

Page 45 – exercise 1

a. Poe's short stories are a mixture of imagination, logic and hallucination.
b. His texts can be grouped into tales of the grotesque and tales of ratiocination.
c. Dupin's approach consisted in looking at the surface of things.
d. The type of mystery that the hardboiled detective investigates is almost the inversion of the tradition of Poe.
e. In hardboiled fiction, the narrator starts with a relatively simple matter which becomes terribly complicated as the story develops.
f. Hammett's stories are far less cerebral than Poe's elaborate and clever puzzles. They are also more "human".

Page 51 – exercise 1

a. party   b. search   c. nail   d. sill
e. rod   f. shutter   g. shrill

Page 51 – exercise 2

a. T
b. F – Both doors facing the corridor were locked from the inside.
c. F – The chimneys were very narrow.
d. T
e. F –There were only two windows and only one of them was nailed shut.
f. T
g. F – Less than two metres from this window there was a lightning rod which ran to the ground.
h. T
i. T

Page 52 – exercise 3

1st – The murderers must have passed through the windows in the back room.
2nd – The murderers must have been really agile.
3rd – Money could not have been the motive for the murders.

Page 52 – exercise 4

Possible answers:
a. Who were the murderers?
b. Where did they come from?
c. What language did they speak?
d. How did they get to be so agile?
e. Why did they kill the two women?

Page 53 – exercise 5

a. Dupin was a brilliant investigator who didn't take long to find out what had really happened. OR Dupin, who was a brilliant investigator, didn't take long to find out what had really happened.
b. The murderer, whose strength must have been remarkable, had almost decapitated Madame L'Espanaye.
c. Dupin searched the flat where he found a lot of interesting things.
d. Dupin had a lot of intuitions which

he told to the narrator.

e. The narrator, who was not as insightful as Dupin, was astonished by his reconstruction of the events.

f. The police inspected the windows which appeared to be locked.

g. Each window had shutters whose width was about one metre.

h. The murderer, who climbed up the wall, must have been of an extraordinary agility.

i. One of the two voices Dupin talked about was extremely shrill.

Page 54 – exercise 1

themselves / strangled / pushed / excessive / took / pull / must / evidence / strength / hair / extremely / pull / half

Page 60 – exercise 1

a. The narrator thought that the murderer must have been a madman.

b. Dupin didn't agree because the murderer spoke a language which contained completely unrecognisable words. Moreover, madmen have human hair while the hair they found in the house seemed to belong to an animal.

c. Dupin got the narrator to wrap the drawing around a cylinder of wood.

d. They concluded that the hand in question was not human.

e. Dupin showed the narrator an article from a book on ethology to prove that the hand belonged to an orang-outang.

f. One of the voices belonged to a Frenchman while the other belonged to the orang-outang.

g. The advertisement stated that an orang-outang had been found and the owner could collect it.

h. Dupin wrote it in order to get the attention of the owner of the orang-murderer.

i. The piece of ribbon proved that the owner of the orang-outang was a sailor.

j. Dupin was expecting a visit from a Maltese sailor.

Page 60 – exercise 2

Open answer.

Page 61 – exercise 3

a. The orang-outang was of great value and the sailor was poor.

b. The orang-outang had been found in the Bois du Boulogne, a very great distance from the scene of the murders.

c. Nobody could think that the animal was responsible for the murders.

d. The sailor thought that if he didn't collect the animal he would attract suspicion.

e. The sailor planned to collect the orang-outang and keep it until the incident had been forgotten.

Page 61 – exercise 4

Open answers.

APES GOING APE

Page 64 – exercise 1

a. T

b. F – King Kong was brought to New York for the entertainment of the masses.

c. T

d. F – They are set in a simian version of the Roman Empire.

e. T

f. F – Romero's film is quite believable.

Page 65 – exercise 1

a. The orang-outang was four or five.

b. The story began on the island of Borneo.

c. The orang-outang was holding a razor in its hand when it ran away from the sailor's place.

d. Adolphe Le Bon was immediately released.

e. The orang-outang was sold to the city zoo.

Page 74 – exercise 1

a. F – He was a very muscular man.

b. F – Dupin didn't want any money at all.

c. T

d. T

e. F – The orang-outang was captured by a friend of his.

f. T

g. T

h. T

i. T

Page 74 – exercise 2

Open answer.

Page 75 – exercise 3

Possible answers:

a. If the sailor hadn't gone to Borneo, he would never have met the orang-outang.

b. If the orang-outang hadn't found the razor, he would have never killed the two women.

c. If the window had been properly locked, the orang-outang wouldn't have been able to get into the house.

d. If Madame L'Espanaye had been fitter, she would have been able to kill the animal.

e. If the orang-outang hadn't escaped from the house, the police would have realised immediately that it was responsible for the murders.

Page 75 – exercise 4-6

Open answers.

PARIS IN THE 1800'S

Page 82 – exercise 1

a. During his reign, three kilometres of quays were built from the Louvre to the Tuileries, Rue de Rivoli was created and the construction of the Arc de Triomphe was begun.

b. Work projects slowed down due to a combination of the kings' lack of ambition and lack of funds.

c. According to doctors, people were supposed to take baths only once a month.

d. The "bouquinistes" were used-books merchants.

e. The department store.

f. Public transport included trams, funicular railways, buses and boats.

g. It turned white under the rain.

h. The Eiffel Tower was built to commemorate the centennial of the French Revolution.

i. The "Belle Époque" started in the 1880's and was brutally interrupted by the First World War.

THE PURLOINED LETTER

Page 94 – exercise 1

Possible answers:

a. It was a windy evening in the autumn of 18__.

b. The narrator and Dupin were friends.

c. For one hour they sat together in profound silence.

d. Dupin was watching the smoke rising from his pipe.

e. The narrator was thinking how fast Dupin had been in solving the Rue Morgue case.

f. Suddenly Monsieur G__ arrived.

Page 94 – exercise 2

Possible answers:

a. If anybody knew that the policeman Monsieur G. told Dupin about the case, he would probably lose his job.

b. If the affair hadn't been so simple, the police wouldn't have been so puzzled.

c. If the document had passed out of the robber's possession, it would have had dramatic consequences.

d. If Dupin hadn't been consulted, the case certainly wouldn't have been solved.

e. If the Minister D. realized the police were looking for the letter, he would disclose its contents.

f. If the minister hadn't often been absent from home, the police wouldn't have been able to search his place.

Page 95 – exercises 3, 4

Open answers.

Page 96 – exercise 5

a. T
b. T
c. F – For each room they spent an entire week searching.
d. F – They also examined the chairs.
e. T
f. T
g. T
h. T
i. F – They also searched the two adjoining houses.
j. T

Page 96 – exercise 6

Open answers.

Page 105 – exercise 1

a. They were occupied in the same manner as before.
b. A very large reward.
c. The Prefect appeared absolutely amazed.
d. The problem with their techniques was that they were not suitable for this case.
e. The minister was also a mathematician and a poet.
f. One shouldn't be either too deep or too superficial in carrying out an investigation.

Page 105 – exercise 2

a. This game is played with marbles.
b. One player holds a number of marbles and asks the other player whether the number is even or odd.
c. If the other player is right, he wins one.
d. On the other hand, if he is wrong, he loses one.
e. In order to win you must first of all observe and measure the intelligence of your opponent.
f. If you think the opponent is a complete idiot, and if you lose the first time answering "odd", you have to answer "odd" the second time.
g. This is because the opponent will think that he will fool you just by switching from even to odd.
h. On the other hand, if your opponent seems to be slightly more intelligent than the first, you should answer "even".
i. This is because this "idiot a degree more intelligent than the first" will probably think that a simple variation would be too predictable and decide to keep the marbles even as before.

Page 106 – exercise 3

a. The Prefect was asked a question about the purloined letter.
b. When the policeman went to see Dupin, the house had been searched several times.
c. A large sum of money is being offered as a reward for finding the letter.
d. To the policeman's surprise, the letter was returned to him by Dupin.
e. The Prefect was surprised that the letter had been found by Dupin.
f. The wrong techniques had been used by the Parisian police to solve the crime.

Page 107 – exercise 4

a. bedpost   b. marbles
c. concealment   d. mathematician
e. puzzled   f. drawer   g. robber
h. perseverance   i. search
j. thickness

Page 107 – exercise 5

Open answer.

Page 108 – exercise 1

head / minister / spectacles / business /
table / books / purloined /
mantlepiece / compartments / tear /
female / letter / appearance /
Prefect / seal

Page 114 – exercise 1

a. T
b. T
c. T
d. F – He went only twice.
e. F – He pretended that he had
forgotten his tobacco.
f. F – Dupin substituted it with a
facsimile and so the minister never
noticed.

Page 114 – exercise 2

Open answers.

Page 115 – exercise 3

Letter A
– The seal was large and black.
– The seal had D's insignia inscribed
on it.
Letter B
– The letter was addressed to D.
– It was written in small female
handwriting.
– The seal was small and red.
– The seal bore the coat of arms of the
Royal Family.
– The letter was addressed to a lady.
– It was written in a large and bold
handwriting.

Page 115 – exercise 4

Open answer.

Page 116 – exercise 5

Open answers.

KEY TO THE EXIT TEST

Focus on the context

a. In 1809 in Boston.
b. John Allan and Frances Keeling
Valentine.
c. He was disappointed because it
was totally ignored by the critics.
d. Some critics consider it to be the
first ever detective story.
e. On October 7, 1849.

Focus on the story

The Murders in the Rue Morgue
1. D   2. B   3. C   4. A   5. C   6. D

The Purloined Letter
7. B   8. D   9. C   10. A   11. D   12. D

2

1. strangled
2. throat
3. witnesses
4. murderer
5. window
6. which
7. therefore
8. neither
9. advertisement
10. who

NOTES

Black Cat English Readers

Level 1
Peter Pan
Zorro!
American Folk Tales
The True Story of Pocahontas
Davy Crockett
Great Expectations NEW!
Rip Van Winkle and The Legend
of Sleepy Hollow NEW!
The Happy Prince and The Selfish
Giant NEW!
The American West NEW!
Halloween Horror NEW!

Level 2
Oliver Twist
King Arthur and his Knights
Oscar Wilde's Short Stories
Robin Hood
British and American
Festivities

Level 3
Alice's Adventures in Wonderland
The Jumping Frog
Hamlet
The Secret Garden
Great English Monarchs and their
Times

Level 4
The £1,000,000 Bank Note
Jane Eyre
Sherlock Holmes Investigates
Gulliver's Travels
The Strange Case of Dr Jekyll
and Mr Hyde
Classic Detective Stories
The Phantom of the Opera
Alien at School
Romeo and Juliet
Treasure Island

Level 5
A Christmas Carol
The Tragedy of Dr Faustus
Washington Square
A Midsummer Night's Dream
American Horror
Much Ado About Nothing
The Canterbury Tales
Dracula
The Last of the Mohicans
The Big Mistake and Other Stories

Level 6
Frankenstein
Pride and Prejudice
Robinson Crusoe
A Tale of Two Cities
The X-Files : Squeeze
Emma NEW!
The Scarlet Letter NEW!
Tess of the d'Urbervilles NEW!
The Murders in the Rue Morgue
and The Purloined Letter NEW!
The Problem of Cell 13 NEW!

BLACK CAT ENGLISH CLUB
Membership Application Form

BLACK CAT ENGLISH CLUB is for those who love English reading and seek for better English to share and learn with fun together.

Benefits offered: *- Membership Card*   *- English learning activities*

                       - Book discount coupon   *- Black Cat English Reward Scheme*

                       - English learning e-forum   *- Surprise gift and more...*

Simply fill out the application form below and fax it back to 2565 1113 or send it back to the address at the back.

Join Now! It's FREE exclusively for readers who have purchased *Black Cat English Readers* !

(Please fill out the form with **BLOCK LETTERS**.)

The title of Black Cat English Reader/book set that you have purchased: _____

English Name: _____ (Surname) _____ (Given Name)

Chinese Name: _____

Address:

▢▢▢▢▢▢▢▢▢▢▢▢▢▢▢▢▢▢▢▢▢▢▢▢▢▢▢▢▢▢▢▢▢▢▢▢▢▢

▢▢▢▢▢▢▢▢▢▢▢▢▢▢▢▢▢▢▢▢▢▢▢▢▢▢▢▢▢▢▢▢▢▢▢▢▢▢

▢▢▢▢▢▢▢▢▢▢▢▢▢▢▢▢▢▢▢▢▢▢▢▢▢▢▢▢▢▢▢▢▢▢▢▢▢▢

Tel: _____ Fax: _____

Email: _____

Sex: ❑ Male   ❑ Female   (Login password for e-forum will be sent to this email address.)

Education Background: ❑ Primary 1-3   ❑ Primary 4-6   ❑ Junior Secondary Education (F1-3)

                             ❑ Senior Secondary Education (F4-5)   ❑ Matriculation

                             ❑ College   ❑ University or above

Age: ❑ 6 - 9   ❑ 10 - 12   ❑ 13 - 15   ❑ 16 - 18   ❑ 19 - 24   ❑ 25 - 34

      ❑ 35 - 44   ❑ 45 - 54   ❑ 55 or above

Occupation: ❑ Student   ❑ Teacher   ❑ White Collar   ❑ Blue Collar

                 ❑ Professional   ❑ Manager   ❑ Business Owner   ❑ Housewife

                 ❑ Others (please specify: _____)

As a member, what would you like **BLACK CAT ENGLISH CLUB** to offer:

     ❑ Member gathering/ party   ❑ English class with native teacher   ❑ English competition

     ❑ Newsletter   ❑ Online sharing   ❑ Book fair

     ❑ Book discount   ❑ Others (please specify: _____)

Other suggestions to **BLACK CAT ENGLISH CLUB**: _____

_____

Please sign here: _____ (Date: _____)

Visit us at Quality English Learning Online http://publish.commercialpress.com.hk/qel

BLACK CAT ENGLISH CLUB
The Commercial Press (Hong Kong) Ltd.
9/F, Eastern Central Plaza,
3 Yiu Hing Road, Shau Kei Wan,
Hong Kong